ELECTRONICS 2/3

Other titles in the Nelson TEC book Series

Engineering Drawing and Materials 1 W. J. Baker
Engineering Science 2 T. Lowe
Engineering and Mechanical Science 3 G. D. Jones
Mathematics 1 J. B. Kettlewell
Construction Technology 3 K. Roberts

Nelson TEC books

ELECTRONICS

C. Kelly

Lecturer in Electronics and Telecommunications,
Somerset College of Arts and Technology, Taunton.

Nelson

For Diane, my wife

Thomas Nelson and Sons Ltd
Nelson House Mayfield Road
Walton-on-Thames Surrey KT12 5PL

P O Box 18123 Nairobi Kenya

116-D JTC Factory Building
Lorong 3 Geylang Square Singapore 1438

Thomas Nelson Australia Pty Ltd
480 La Trobe Street Melbourne Victoria 3000

Nelson Canada Ltd
1120 Birchmount Road Scarborough Ontario M1K 5G4

Thomas Nelson (Hong Kong) Ltd
Watson Estate Block A 13 Floor
Watson Road Causeway Bay Hong Kong

Thomas Nelson (Nigeria) Ltd
8 Ilupeju Bypass PMB 21303 Ikeja Lagos

First published 1981

Reprinted 1982

ISBN 0-17-741123-6
NCN 450-5845-1

Filmset in Hong Kong by Asco Trade Typesetting Ltd
Printed and bound in Hong Kong

Preface

This textbook is intended for those studying Technician Education Council (TEC) courses in electronics and covers Level Two and Level Three topics. It closely follows the TEC standard units of Electronics Two (U76/010) and Electronics Three (U76/009) and should meet the requirements of college devised units.

Although it is split into two parts, (entitled Electronics Level Two and Electronics Level Three), there is some overlap, the first part providing many of the basic principles of electronics necessary for a full understanding of Level Three.

Each chapter is designed to make learning as easy as possible. It begins with a brief description of its aims, and is interspersed with Reviews followed by questions which are designed to help the student pause, recall what has just been covered and ensure that s/he understands it. The text is therefore divided into easily manageable parts. At the end of each chapter there is a series of multiple choice and short answer self-assessment questions which test the reader's learning achievement.

Contents

Part A: Electronics Level 2

6 Cathode ray oscilloscope

7 Oscillators

8 Digital electronics

Part B : Electronics Level 3

9 Field effect transistor

10 Amplifiers

11 Electrical noise

12 Feedback amplifiers

13 Stabilised power supplies

14 Pulse generators and pulse-shaping networks

15 Integrated circuits

Part A: Electronics Level 2

1 Introduction to electronics

This chapter describes the properties and manufacture of the materials of which most active electronic devices are made today. The aims of this chapter are as follows:

- to describe the difference between conductors, insulators and semiconductors
- to provide an appreciation of the term semiconductors
- to provide a simplified explanation of the atomic theory of matter
- to provide an understanding of the properties and processing of materials commonly used in active electronic devices (such as junction diodes, transistors and integrated circuits).

1.1 What is electronics?

The atoms within a material contain protons and **electrons**. **Electricity** is a description of the form of energy present in protons and electrons. Strictly, **electronics** describes the science and technology dealing with the behaviour of electrons.

It is generally accepted that electronics and in particular **electronic engineering** has a broader description than above: It is concerned with devices, materials, components and systems which process and control the flow of small quantities of electrical energy.

To appreciate the term 'electronics' it helps to associate it with some of the familiar systems which it embraces, such as radio, television, hi-fi amplifiers, computers and so on. However, the study of electronics covers the devices and components which are contained in these systems such as transistors, integrated circuits, resistors, capacitors and so on. Furthermore, electronics is concerned with the materials from which these devices and components are made.

Electronic engineering is considered to be the fastest changing technology today. The development of small computers and their central processors called **microprocessors** is influencing the technological advance of present day society in a revolutionary way.

1.2 Studying electronics

Although it is possible to obtain an understanding of the behaviour of electronic systems by a description of their characteristics and purpose this is a superficial approach to the study of electronics.

In order to appreciate electronics and electronic engineering properly it is necessary first to understand the properties of the materials used in electronic devices and second to understand how these materials can be used to make the devices. Only then can a real understanding of the design of electronic systems be appreciated. For example, the main active components in a hi-fi amplifier could be **transistors** which in themselves are crude amplifying devices. In order to understand how transistors amplify it is necessary to appreciate the properties of the materials of which a transistor is constructed.

1.3 Conductors, semiconductors and insulators

An electric current consists of a flow of electric charge (a displacement of electrons or protons).

The conduction of an electric current through a material depends upon the atomic structure of the material. In general, the number of 'free' electrons on the outside of atoms of a material gives an indication of how easily a current can flow through it.

Certain materials such as metals are good **conductors**, others such as rubber, are very bad conductors – called **insulators**. There is a range of materials between the best conductors and the best insulators.

Semiconductors are a class of materials which have special properties and are processed in such a way that their **conductivity** can be modified.

The term semiconductor is also used to describe

the electronic devices made of the semiconducting materials such as junction diodes, transistors and integrated circuits.

Before the action and operation of these devices can be fully understood, a knowledge of the properties of semiconductor materials is required. First, we study these materials in their pure state, in which they are known as **intrinsic** semiconductors (see Sections 1.5 to 1.8). Semiconductors adapted for the purpose of electronics are known as **extrinsic** semiconductors (see Sections 1.9 to 1.12).

Semiconducting materials are able to exhibit the properties of both conductors and insulators. They are true to their name: At room temperature their conducting properties lie somewhere between the considered ideals of good conductors and insulators. Fig. 1.1 compares the resistivities of a selection of conductors, semiconductors and insulators.

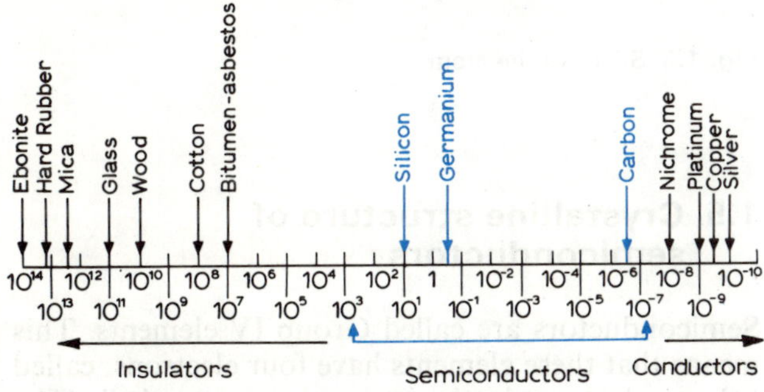

Fig. 1.1 Comparison of resistivities (Ω–cm)

They bear a similarity to insulators in that they have a negative temperature coefficient of resistance (as the temperature increases the resistance decreases), whereas conductors generally have a positive temperature coefficient.

A metallic crystalline element called germanium (symbol Ge) is a semiconducting material commonly used in electronics. It was the first semiconductor used successfully on a large scale for making electronic devices.

Silicon (symbol Si), a non-metallic crystalline element, is the most common semiconducting material used these days. It has superior qualities, such as withstanding high temperatures, when compared with germanium.

Review

Check your knowledge and understanding. You should:

a Appreciate the meaning of the term 'electronics'.

b Appreciate the difference between conductors, semiconductors and insulators.

c Know that two common semiconductor materials are germanium and silicon.

Exercise

1.1 Identify the following statements as true or false:

a The resistivity of a semiconductor is greater than that of an insulator.

b The resistivity of a semiconductor is greater than that of a conductor.

c Semiconductors have a positive temperature coefficient of resistance.

1.2 Identify which of the following are the common semiconductors used in electronics:

a phosphorus, **b** silicon,
c indium arsenide, **d** boron,
e germanium, **f** indium,
g copper oxide.

1.4 Simplified atomic theory of matter

All matter can be subdivided into smaller and smaller parts until eventually the **atom** is encountered. Almost one hundred different types of naturally occurring atoms have been identified and these are called **elements**. Each element has been named, many are familiar to you, e.g. hydrogen, carbon, lead, and so on.

The atom itself can be subdivided into many more components, sub-atomic particles, of which only two are of real concern to the study of electronics. These are the **electron** and the **proton** (see Table 1.1 for details).

Our interest in electrons and protons is because they each possesses a special property called an **electric charge**. Protons and electrics have opposite charges. The proton is said to have a **positive** charge, and the electron a **negative** charge. A fundamental law governing charged particles is that two like charged particles (both positive or both negative) repel each other; two unlike charged particles (one positive and one negative) attract each other.

The charge on the proton and electron differ only in their polarity or sign representation. The magnitude of each is the same and is measured as 1.602×10^{-19} coulombs, the value of which is designated by the symbol e. (It is much more convenient to represent the magnitude of electron charges in e rather than in coulombs, for example the charge on two electrons is $-2e$.)

Protons are the heavier of the two particles (Table 1.1) and are bound by very powerful forces in the

Table 1.1 The three basic components of the atom

	proton	electron	neutron
mass (kg)	1.6725×10^{-27}	9.109×10^{-31}	1.673×10^{-27}
charge	$+e$	$-e$	electrically neutral

$e = 1.602 \times 10^{-19}$ C

centre of the atom (the nucleus). Protons are very difficult to remove from the nucleus, and for our purpose are considered to be fixed permanently to particular atoms. A third type of sub-atomic particle, also fixed in the nucleus, is the neutron, but as they are electrically neutral (they carry no charge) they are of little significance to electronics.

Electrons are much lighter particles than protons and are said to revolve around the nucleus at fixed distances called shells. The reason for describing the path of an electron as a shell is illustrated in Fig. 1.2. The plane of the orbit itself rotates, the electron path describing the surface of an imaginary sphere.

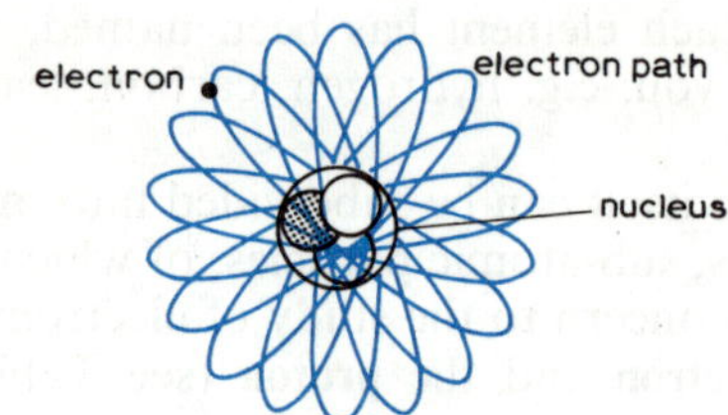

Fig. 1.2 The atom

Normally, an atom has the same number of electrons as protons, the total negative and positive charges balance, so the complete atom is considered electrically neutral. If an atom loses an electron, the protons outnumber the electrons by one; therefore the overall charge is positive by an amount $+e$. The atom is then called a **positive ion**. Occasionally electrons outnumber protons, and such atoms are called **negative ions**.

The number and combination of the sub-atomic particles (electrons, protons and neutrons) in any element determines the distinct properties of that element. For example, consider the simplest element, hydrogen. An atom of hydrogen has a nucleus of one proton, one electron completes the atom. Helium is another simple element. It has a nucleus normally comprising two protons and two neutrons. Two electrons complete the atom.

The electrons of higher order atoms (those with large numbers of electrons and protons) cannot all reside in only one shell. In fact, the electrons distribute themselves in a number of shells (see Fig. 1.3) which are labelled K, L, M, N, O, P, Q from the innermost out. The maximum number of electrons accommodated by each shell is shown at the top of Table 1.2.

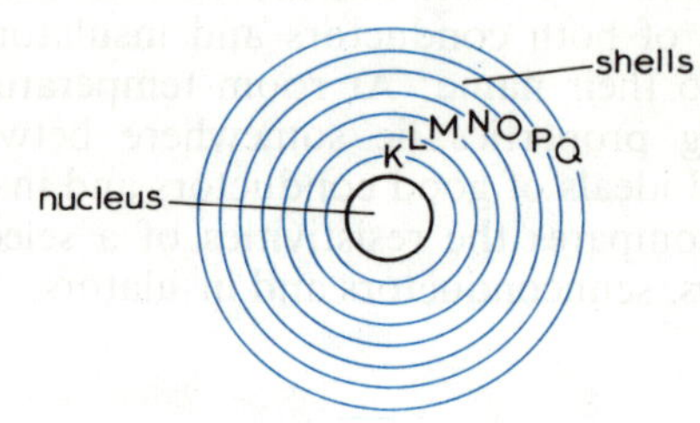

Fig. 1.3 Shells of the atom

1.5 Crystalline structure of semiconductors

Semiconductors are called Group IV elements. This means that these elements have four electrons, called **valence electrons** in the incomplete outer shell. (The occurrence of four valence electrons is sometimes referred to as **tetravalency** – the prefix 'tetra' meaning four.) The electron distribution of silicon and germanium are shown in Table 1.2.

Table 1.2 Important elements of Groups III, IV and V (Valencies shown in bold)

Element		Electron distribution in shells K	L	M	N	O	P	Q
	Maximum →	2	8	18	32	50	72	98
Arsenic	Donors Group V	2	8	18	**5**			
Phosphorous		2	8	**5**				
Antimony		2	8	18	18	**5**		
Germanium	Semiconductors Group IV	2	8	18	**4**			
Silicon		2	8	**4**				
Gallium	Acceptors Group III	2	8	18	**3**			
Aluminium		2	8	**3**				
Boron		2	**3**					
Indium		2	8	18	18	**3**		

Pure semiconductor materials have a crystalline structure, meaning that all the atoms are arranged in an orderly repeated pattern. This arrangement of atoms is called a **crystal lattice**. Any one atom is positioned such that it has four equidistant neighbours (Fig. 1.4a). For clarity, this situation is shown in a two dimensional diagram (Fig. 1.4b).

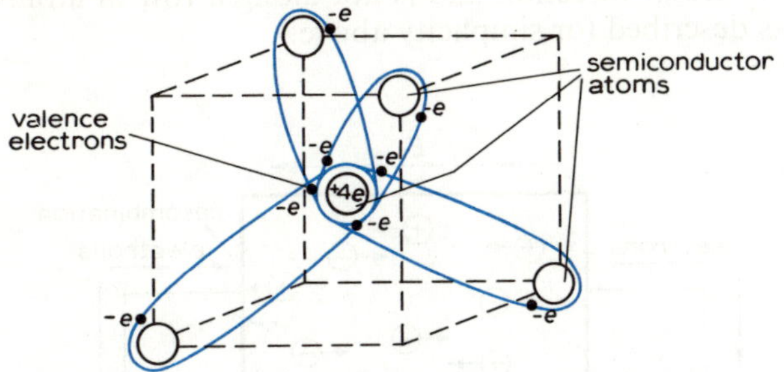

Fig. 1.4a The cubic structures of semiconductor atoms

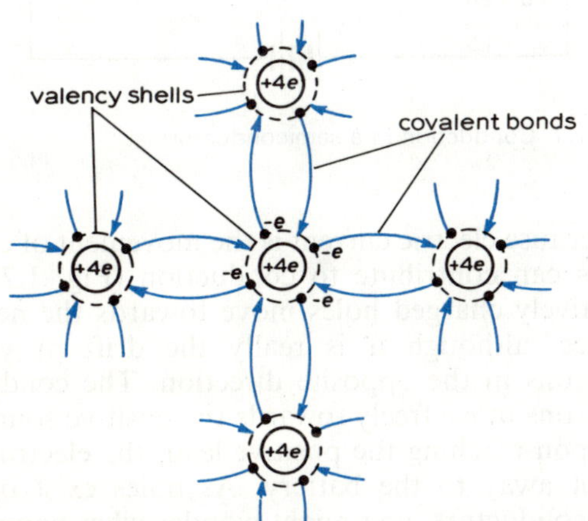

Fig. 1.4b Two-dimensional representation of cubic structure

The four valence electrons of each atom are shared with the four nearest neighbours, one electron with each. An orbit is established between neighbouring atoms, each atom contributing to four orbits. Although the electrons are continually interchanged between atoms, at any instant in time each atom has a total of four valence electrons to preserve the balance of charge. These orbits are called **covalent bonds**, and each requires two electrons (one from each neighbouring atom) to be complete.

The sharing of valence electrons enables atoms to be rigidly bonded together. Also by this means is achieved a much tighter attachment of the valence electrons to the atom than with a single semiconductor atom bearing an incomplete outer shell.

As no semiconductor is entirely pure the foregoing description of a semiconductor material is simplified and idealised. Small traces of other elements always occur in semiconductors. These impurities very slightly disrupt the perfect symmetry of the pure crystalline material.

1.6 Charge carriers in semiconductors

The crystal bonding of semiconductor atoms might at first appear to leave no electrons available for conduction. This is true at low temperatures when all the covalent bonds are complete. However, at

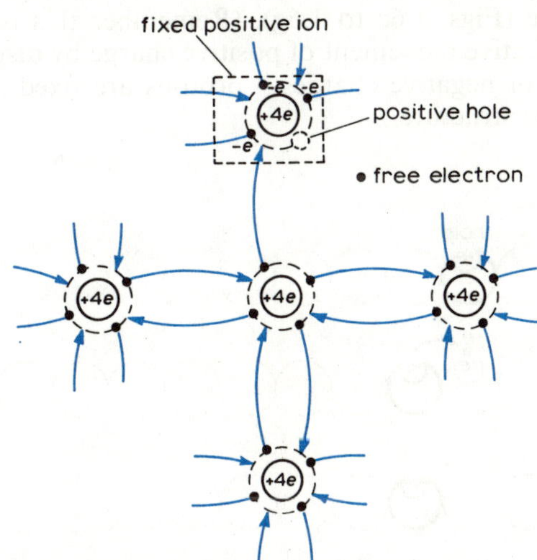

Fig. 1.5 An electron-hole pair within a semiconductor

higher temperatures, a small number of the valence electrons attain sufficient thermal energy to break away from their bonds (Fig. 1.5). Such electrons then become free to wander through the lattice and are available for conduction.

In a group of atoms which has lost an electron (Fig. 1.5) one bond is incomplete. In this group the protons outnumber the remaining electrons by one. Therefore, there is a localised net positive charge of $+e$. As this is a direct result of an electron vacating its bond, it is called a **hole**.

Because free electrons and holes are generated simultaneously, they are called **electron-hole pairs**.

Intrinsic semiconductors have the same number of holes as conduction electrons.

Free electrons are able to move from hole to hole. A free electron combining with an available hole is called **recombination**. Recombination and generation of fresh pairs occurs continuously but it can be assumed that a piece of intrinsic semiconductor has a constant number of electron-hole pairs, although not the same ones.

1.7 Conduction in semiconductors

Consider ten atoms in a row in a semiconductor of which one has a hole (Fig. 1.6a). If the hole attracts an electron from the atom adjacent it appears as if the hole has moved one atom along (Fig. 1.6b). If this process continues with the next atom in the row and so on, the effect is one of a hole moving along the row of atoms; an effective movement of positive charge (Figs. 1.6c to 1.6g). (Remember this is only an effective movement of positive charge by displacement of negative charges – protons are fixed in the atomic structure.)

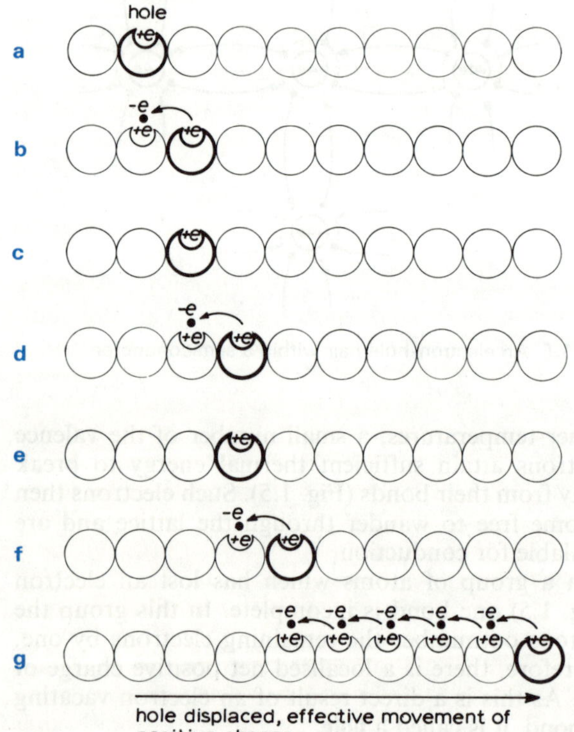

Fig. 1.6a–1.6g

To enable significant charge displacement, either described as a movement of electrons or holes, in a semiconductor an electric field, supplied externally, must exist.

Electric current is a movement of charge and occurs in a semiconductor, as described above, if the external electric field is provided by a battery (Fig. 1.7). The flow of charge is a general movement in a particular direction and is not along a row of atoms as described for simplicity above.

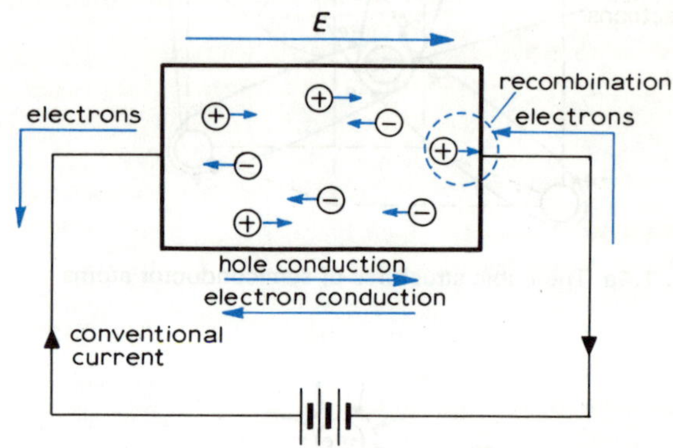

Fig. 1.7 Conduction in a semiconductor

Because electric current is the movement of charge, holes can contribute to conduction (Fig. 1.7). The positively charged holes move towards the negative source, although it is really the drift of valence electrons in the opposite direction. The conduction electrons move freely towards the positive source.

Upon reaching the positive lead, the electrons are swept away to the battery. As holes exist only in semiconductors, you might wonder what happens to them as they reach the negative lead. In fact, fresh electrons are drawn from the negative source of the battery to recombine with the holes. As new electron-hole pairs are continually being generated within the material, current flow is maintained as long as the battery is connected.

Therefore there are two types of current in a semiconductor; electron current, I_e, and hole current, I_h.

As current is defined as the total charge moved in a given time, the two types of current are added. Hence,

$$\text{Total current } I = I_e + I_h$$

1.8 Effect of temperature on intrinsic conduction

The number of charge carriers in a semiconductor are temperature dependent. As the temperature increases, a greater number of valence electrons attain sufficient thermal energy to escape their covalent bonds. Hence, more electron-hole pairs are generated. As the number of electron-hole pairs increases with rising temperature, then so does the conductivity. The reason for this is that at higher temperatures, more charge carriers are available for conduction for a given e.m.f.

The conductivity of silicon increases approximately 8 percent per degree centigrade increase in temperature. The conductivity of germanium increases approximately 6 percent per degree centigrade.

Review

Check your knowledge and understanding. You should:

a Understand the basic structure of intrinsic semiconductors.
b Appreciate the actions of holes and electrons.
c Understand intrinsic conduction by holes and electrons.
d Know how a change in temperature affects intrinsic conduction in a semiconductor.

Exercise

1.3 Determine the expected shell distribution of elements with
a 14 electrons **b** 33 electrons **c** 13 electrons
Identify to which group of elements each belongs, e.g. an electron with 31 electrons
K = 2, L = 8, M = 18, N = 3 electrons
Valency of 3, therefore this is a Group III element.

1.4 Identify the following statements as true or false:
a A semiconductor has a crystalline structure.
b Mobile charge carriers are produced thermally.
c Electrons and holes are mobile charge carriers.
d An intrinsic semiconductor has more conduction electrons than holes.
e An increase in temperature increases the resistivity of an intrinsic semiconductor.

1.9 Extrinsic semiconductors

At this stage we have considered only intrinsic or unmodified semiconductor material. Intrinsic semiconductors have limited application to electronics, the main one being temperature variable resistors called thermistors. Modifications to their basic composition are necessary before they can be used for semiconductor devices such as diodes and transistors.

For the construction of semiconductor devices more charge carriers (holes and electrons) independent of temperature are required. These can be obtained by adding to the semiconductor specific amounts of special impurities during manufacture of the material. This can be done during the molten state or added to the solid at a later time. The process is called **impurity doping** and the special impurities called **dopants**.

Depending on the chemical nature of the dopant, the same type of semiconductor can be modified in two distinct ways. One is the creation of a greater number of conduction electrons than holes within the material called ***n*-type** (negative) semiconductor. The second gives a greater number of holes, making a ***p*-type** (positive) semiconductor. *n* and *p*-type are collectively called **extrinsic semiconductors**.

1.10 *n*-type semiconductors

To make the *n*-type extrinsic semiconductor, Group V dopants are used (Table 1.2). The numeral V means that atoms of elements within this group have five valence electrons.

Group V dopants are introduced into the semiconductor such that each impurity atom fits neatly into the crystal lattice, occupying the space that would normally be filled by a semiconductor atom (Fig. 1.8a). Each impurity atom must then take part in the covalent bonding described in Section 1.5.

Semiconductors are generally Group IV elements, each atom has four valence electrons each contributing to a covalent bond, four bonds per atom for a stable crystal structure.

The Group V dopant readily contributes four of its five valence electrons for this purpose, but one electron remains surplus to the bonding (Fig. 1.8a).

Being excluded from the tight covalent bonding, the spare electron is loosely attached to the impurity atom. At room temperature it can attain sufficient thermal energy to detach itself from the group and become a free electron available for conduction.

The spare electron of each impurity atom wanders off through the lattice and leaves behind an excess positive charge of $+e$ (the balancing proton in the impurity nucleus) which is permanently fixed in that position (Fig. 1.8b). These fixed **positive ions** are not to be confused with holes as each bond is complete.

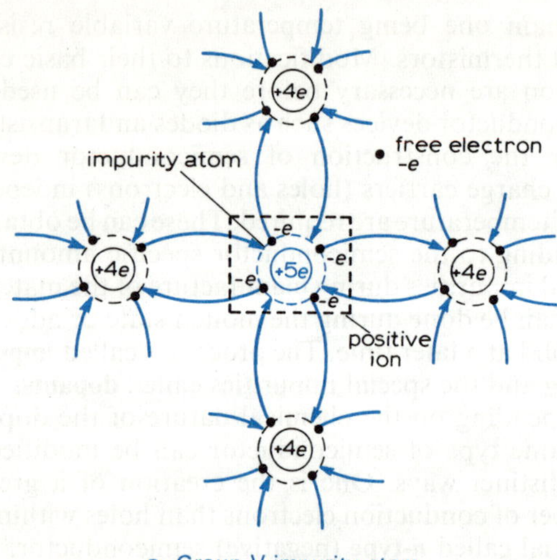

a Group V impurity atom

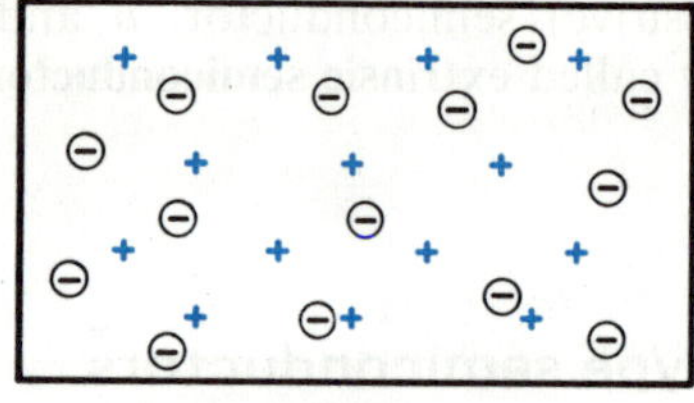

b Fixed and mobile charges in *n*-type

Fig. 1.8 *n*-type semiconductor

Every Group V impurity atom added donates four electrons to the covalent bonding and one to the number of conduction electrons. Group V impurities are therefore called **donors**. Typical donors are arsenic, phosphorous and antimony (see Table 1.2).

Because most of the conduction is now by negative charge carriers, the material is called *n*-type. There are still relatively few naturally occurring holes as with intrinsic materials.

In *n*-type material the **majority carriers** are *electrons* and the **minority carriers** are *holes*.

However, throughout the *n*-type material there is still a balance between electrons and protons, so the overall electric charge remains neutral. Doping with Group V donors essentially enables more of the electrons to be easily freed from the impurity atoms and made available for conduction.

Doping densities are in the range of 1 in 10^3 to 1 in 10^8 depending on the materials used and the desired application. A general doping density of 1 in 10^5 means that one impurity atom is added for every 100 000 semiconductor atoms. This may initially seem to be an extremely low number, but when you consider that a chip of silicon one cubic centimetre contains 5×10^{22} atoms, the number of free electrons created by doping is typically

$$\frac{5 \times 10^{22}}{10^5} = 5 \times 10^{17}$$

This is sufficient to increase considerably the conductivity.

1.11 *p*-type semiconductors

The extrinsic semiconductor called *p*-type is manufactured by adding Group III impurities to the pure crystal. Normally, atoms of Group III elements each have three valence electrons.

Each impurity atom fits in the crystal lattice in the place of a semiconductor atom and must take part in the covalent bonding (Fig. 1.9a). As there are only three valence electrons available per impurity atom, one bond is incomplete.

At this stage, a hole has not yet been produced as the whole region has a balance of charge. Only when an electron from a neighbouring bond fills the incomplete impurity bond is a hole left behind (Fig. 1.9b). (It may seem illogical that an extra electron should be attracted to the impurity atom, and so it must be assumed that in certain circumstances the attraction to complete the bond is greater than the electrostatic attraction of the hole.)

The bond which includes the impurity atom is said to accept an extra electron. So dopants used in the

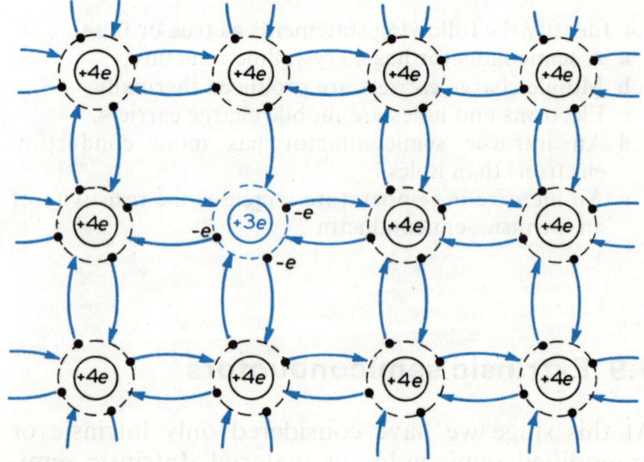

Fig. 1.9a Group III impurity in semiconductor

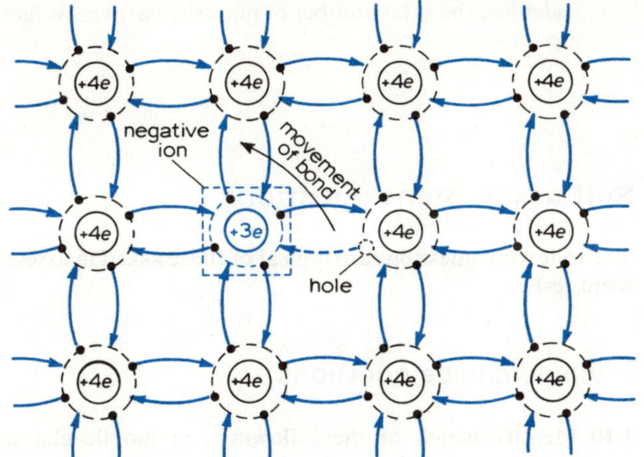

Fig. 1.9b Hole created by presence of Group III impurity

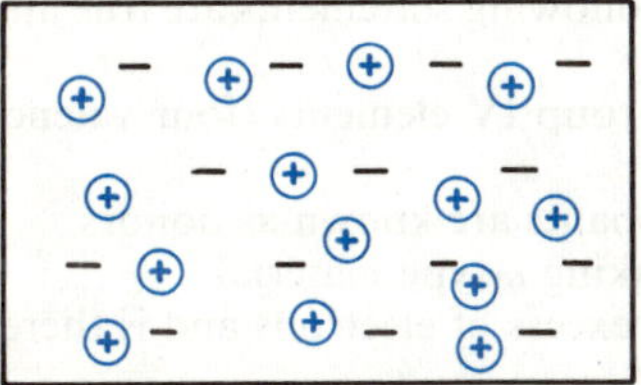

Fig. 1.9c *p*-type semiconductor

manufacture of *p*-type semiconductors are called **acceptors**. Every impurity atom then becomes a **negative ion** permanently fixed in the lattice (Fig. 1.9c). The hole is a charge carrier; the ion is a stationary negative charge.

In addition to these artificially created holes, the electron-hole pairs from thermal agitation are also present as in intrinsic semiconductors. The material is now called *p*-type because most of the conduction is by the positively charged holes.

In *p*-type material the *majority carriers* are *holes* and the *minority carriers* are *electrons*.

Typical acceptor dopants are gallium, aluminium, boron and indium.

1.12 Processing semiconductor materials

Natural semiconductor materials are far from pure, containing many unwanted impurities. Silicon can be made about 98% pure by smelting the common forms of silicon oxide (e.g. sand and quartz) with carbon in a furnace. A much greater purity is required for use in semiconductor devices.

Two examples of refining semiconductors are:

1 Growing a crystal (Fig. 1.10a) involves touching the surface of the molten semiconductor with a small piece of very pure semiconductor of the same kind, called a seed crystal. As the seed is drawn slowly away from the melt and simultaneously rotated, the semiconductor atoms solidify around the seed into a very pure crystal. This process takes place in an atmosphere of inert gas such as helium or neon to prevent contamination.

2 Zone refining a rod of semiconductor (Fig. 1.10b) is done by heating a small localised section by a radio frequency (RF) heating coil. (A frequency of 13.5 MHz typically is used for heating by induction.) The coil is swept down the bar slowly and repeatedly. The impurities tend to gather in the molten section and a pure crystal is left behind. The end section containing the impurities is then cut off and discarded.

Practical methods of adding dopant impurities are briefly discribed in subsequent sections when relevant to the desired application.

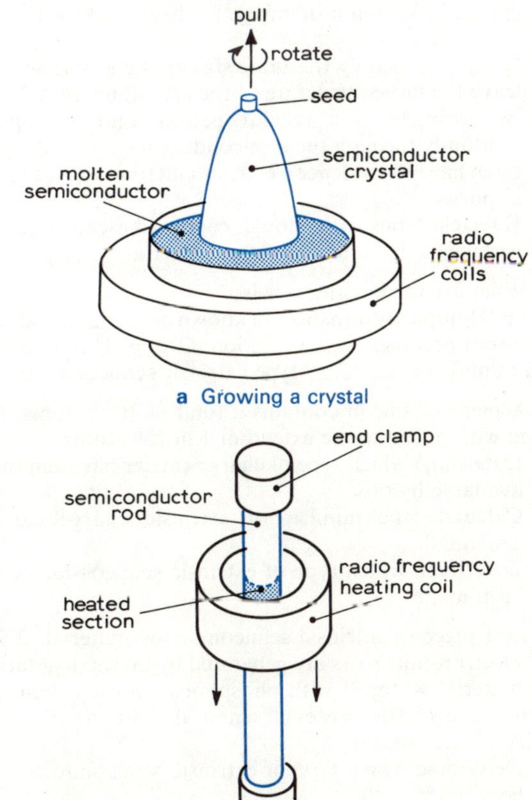

Fig. 1.10 b Zone refining

Review

Check your knowledge and understanding. You should:

a Understand the basic structure of *n*-type and *p*-type semiconductors.

b Know the dopants used for making *n*-type and *p*-type semiconductors.

c Appreciate that electrical conduction is a movement of electrons in *n*-type material and an 'apparent' movement of holes in *p*-type.

Exercise

1.5 Identify which of the following group of facts refers to a Group III element:

a Three valence electrons; used for *p*-type semiconductors; known as an acceptor.

b Five valence electrons; used for *n*-type semiconductors; donor.

c Three valence electrons; used for *n*-type semiconductors; acceptor.

d Five valence electrons; used for *p*-type semiconductors; donor.

1.6 Identify which of the properties below describes an element used as a donor impurity. Each atom of the element must:

a Be approximately the same size as the atoms so as to leave the lattice structure of the crystal unaltered.

b Be arranged in a regular pattern and not spread randomly through the semiconductor.

c Each have five valence electrons and thus act as a donor impurity.

d Have electrons with enough energy to escape from the surface of the semiconducting material.

1.7 What are the missing words:
Group III impurity dopants are known as ________ because each atom becomes a ________ ion. Group III dopants are used in making ________ -type extrinsic semiconductor.

1.8 A piece of silicon contains a total of 10^{22} atoms. It is doped with boron to the extend of 1 in 10^6 atoms.

a Determine which type of charge carrier has been made available by this.

b Calculate the number of extrinsic charge carriers generated.

c Determine which type of extrinsic semiconductor has been made.

1.9 In a piece of intrinsic semiconductor material, 2.5×10^{14} electron-hole pairs are generated by thermal agitation. The material is doped with phosphorus to the extent of 1 in 10^3 atoms. The piece of material contains 5×10^{21} semiconductor atoms.

a Determine which type of extrinsic semiconductor has been made.

b Calculate the total number of majority carriers. Which are they?

c Calculate the total number of minority carriers. Which are they?

Self assessment questions

The following questions are typical of those asked in assessment tests.

Multiple choice questions

1.10 Identify which of the following are mobile charge carriers in a semiconductor.

a negative ions
b conduction electrons
c holes
d positive ions
e valence electrons
f impurity atoms

1.11 Identify which of the following statements are true and which are false.

a Semiconductors are Group IV elements (four valence electrons).

b Group III impurity dopants are known as donors.

c Donors are used in making *n*-type material.

d *n*-type material has an excess of electrons and is therefore negatively charged.

1.12 Identify what effect the addition of dopant impurities has on the conductivity of a semiconductor.

a conductivity is unaffected
b conductivity is increased
c conductivity is decreased
d the conductivity is increased or decreased depending on the type of dopant used

1.13 Identify which of the following correctly completes the statement. Impurity atoms added to pure silicon to form a *p*-type semiconductor must:

a each have three valence electrons and act as an acceptor impurity
b be arranged in a regular pattern and not spread randomly through the lattice
c each have five valence electrons and act as a donor
d be approximately the same size as silicon atoms

Short answer questions

1.14 Briefly describe the structure of intrinsic semiconductor materials.

1.15 Describe how electron-hole pairs are generated by heat within an intrinsic semiconductor. Explain how these mobile charge carriers contribute to conduction.

1.16 Describe how *n*-type and *p*-type semiconductors are made by the addition of dopant impurities, identifying the impurity groups required for each.

2 The junction diode

This chapter describes the *p-n* junctions, its characteristics, manufacture and applications. The aims of this chapter are as follows:

- to describe the *p-n* junction
- to provide an appreciation and understanding of the conducting properties of the *p-n* junction
- to describe the characteristics of semiconductor diodes
- to provide a comparison between silicon and germanium diodes
- to provide an appreciation of diode types and their application
- to describe the manufacture of the *p-n* junction
- to describe the particular applications of the semiconductor diode for rectification and voltage stabilisation.

2.1 Introduction

Conducting and semiconducting materials when connected to a source of e.m.f. carry the same magnitude of current whichever way they are connected to the source. Reversing the polarity of the e.m.f. only causes a reversal of the direction of current flow.

Combinations of different types of semiconductors do not exhibit this bi-directional property. In particular, *p*-type and *n*-type materials connected together have the property that current flows through the combination more easily in one direction than the other. That is, the combination, known as a *p-n* junction, has a much higher conductivity when connected one way across a source of e.m.f. than when the polarity is reversed.

Although the term diode describes any two-electrode device it is usually associated with devices which exhibit the property of only allowing an electric current to pass in one direction through them.

The *p-n* junction diode is a very useful and widely used electronic component. In addition, its principle is important in the understanding of the operation of other semiconductor devices.

2.2 Abrupt *p-n* junctions

Fig. 2.1 shows a two-dimensional diagrammatic representation of a *p-n* junction. For convenience mobile charge carriers are shown encircled (*p*-type: mobile holes, ⊕; fixed acceptor ions −, and *n*-type: mobile electrons, ⊖; fixed donor ions +).

The abrupt *p-n* junction is formed by manufacturing a *p*-type and *n*-type semiconductor region side by side (manufacturing processes are described in Section 2.9).

At manufacture mobile charge carriers diffuse across the junction. Combination of these holes and electrons takes place leaving a thin layer at the junction which is depleted of free electrons and free holes. This layer is called the **depletion layer** (Fig. 2.1b and c).

Although the depletion layer has lost its mobile charge carriers, the fixed impurity ions are left uncovered. Equilibrium is reached where further diffusion of carriers is prevented by the electrostatic repulsion of these fixed ions. The uncovered negative ions of the *p*-type prevent the mobile electrons of the *n*-type from crossing the depletion layer; the uncovered positive ions of the *n*-type prevent the holes of the *p*-type from crossing the depletion layer.

The ions of the depletion layer give rise to an electric field of value E acting over the thickness of the depletion layer d. This is the same as having a potential difference or voltage V_b across the junction which opposes any further diffusion (Fig. 2.1d). This small potential difference is known as a *potential barrier* where

$$V_b = E \times d \text{ volts} \qquad (2.1)$$

The value of V_b depends upon the original type of semiconductor material and the extent to which it is doped.

Typically, $V_b = 0.6$ to 0.7 volt for silicon
$V_b = 0.15$ to 0.2 volt for germanium.

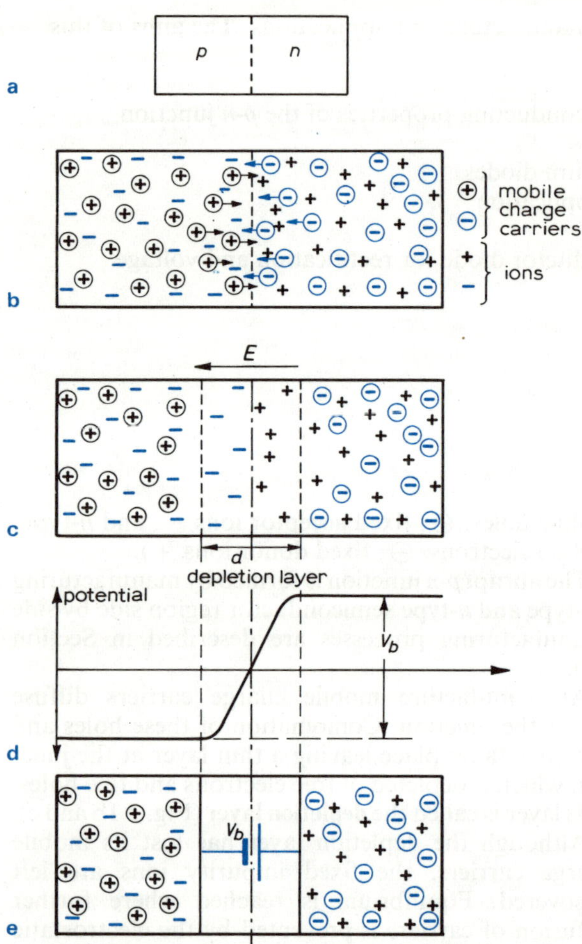

Fig. 2.1 The formation of a *p-n* junction

The potential barrier can be represented by a battery sign as shown in Fig. 2.1e. It must be remembered, though, that this is not a true battery and therefore cannot be connected to supply current to an external circuit as there are no surplus mobile charges provided. The voltage V_b is only representative of the amount of energy required by charge carriers to surmount the barrier. At normal temperatures, very few carriers attain the required energy and so there is no further diffusion.

Although the potential barrier opposes the movement of majority carriers across the junction, also it assists the transition of minority carriers (holes in *n*-type; electrons in *p*-type).

2.3 Reverse biasing the *p-n* junction

The size of the potential barrier can be modified by the connection of an external voltage. To increase the potential barrier, an applied voltage is connected with the positive to the *n*-type and the negative to the *p*-type (Fig. 2.2a). In this configuration, the holes in the *p*-region and electrons in the *n*-region are drawn away from the junction. The thickness of the depletion layer is then increased, thus raising the potential barrier to balance the influence of the applied voltage. The potential barrier is increased to a value $V_b + V$ (Fig. 2.2b).

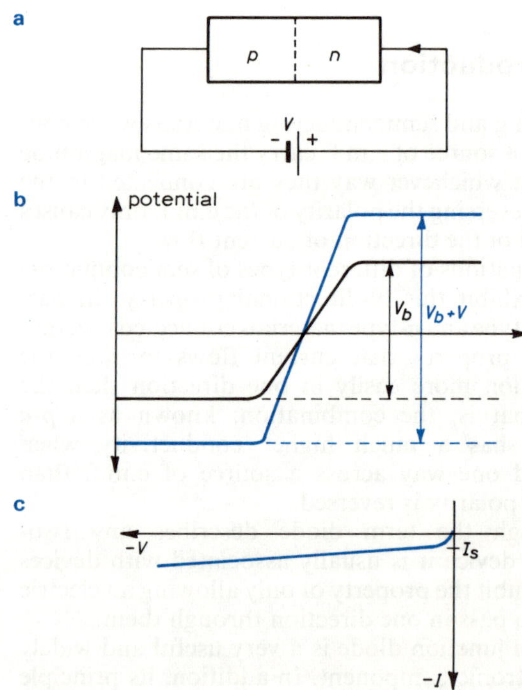

Fig. 2.2 Reverse-biased *p-n* junction

This connection of applied voltage is called **reverse biasing** the junction and inhibits majority carriers from crossing the depletion layer.

Because the potential barrier assists the crossing of the junction by minority carriers, a very small

minority current or **reverse saturation current** I_s flows in the external circuit when the junction is reverse biased. This current is constant over a wide range of applied voltage (Fig. 2.2c). It is normally of the order of micro-amperes for a silicon diode, milli-amperes for a germanium diode, although the actual value varies from diode to diode.

2.4 Forward biasing the *p-n* junction

By **forward biasing** the *p-n* junction with an external voltage V (positive to the *p*-region; negative to the *n*-region as in Fig. 2.3a), the value of the potential barrier is reduced to $V_b - V$ (Fig. 2.3b).

With a reduced potential barrier, more majority carriers have the energy required to pass through the depletion layer. When the value of V exceeds V_b, the barrier is completely overcome. The majority carriers are then forced across the junction by the applied voltage and a substantial forward current I_f flows in the external circuit.

Fig. 2.3c shows how the current initially increases slowly with increasing applied voltage, until the potential barrier is exceeded and then the current increases rapidly. The total current I_f flowing in the external circuit is the addition of the hole and electron currents. (see Section 1.7).

The forward and reverse characteristics can be plotted on the same axes (Fig. 2.4). Note the difference in scales between positive and negative values.

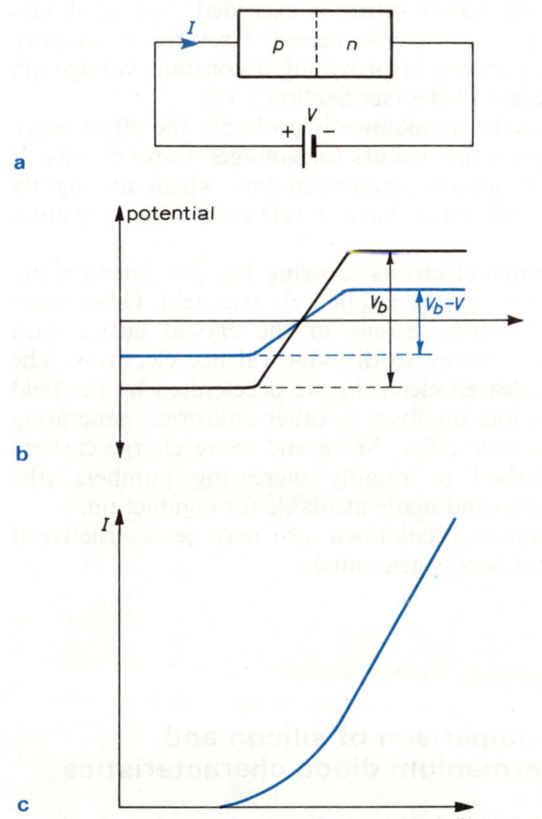

Fig. 2.3 Forward-biased *p-n* junction

2.5 Reverse and forward resistance

The characteristic of the diode which is so useful in electronics, is that it conducts very well in only one direction. Ideally, a diode should behave as a **short circuit** (have zero resistance) when forward biased, and as an **open circuit** (have infinite resistance) when reverse biased. In practice, useful electronic devices have characteristics far from the ideal.

To obtain some idea about the performance of a diode its forward and reverse resistances are specified.

Because the reverse current is very small for a given applied voltage, the **reverse resistance** must be high. Taking any point on the reverse characteristic of Fig. 2.5 and dividing the voltage by the current at this point gives a d.c. resistance. Because the d.c. values differ along the characteristic they are of little

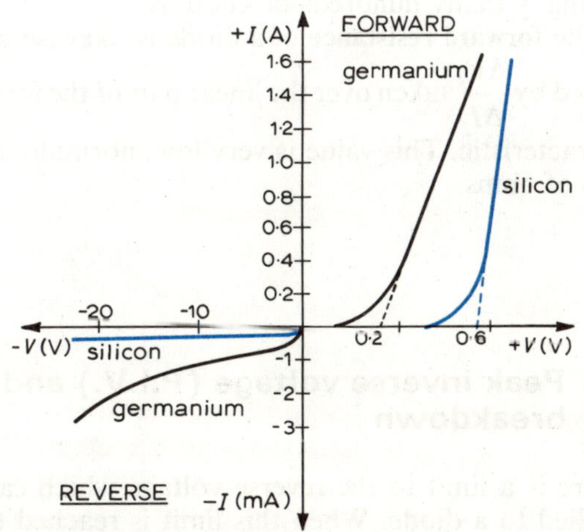

Fig. 2.4 Comparison of silicon and germanium diode characteristics

use. Diodes are largely employed using a.c. or time-varying voltage, so a constant value of resistance is defined for these conditions. By dividing the change in reverse voltage (denoted ΔV_r) by the change in reverse current (ΔI_r) we have a constant value of reverse resistance (see Fig. 2.5).

$$\text{reverse resistance} = \frac{\Delta V_r}{\Delta I_r} \text{ ohm} \qquad (2.2)$$

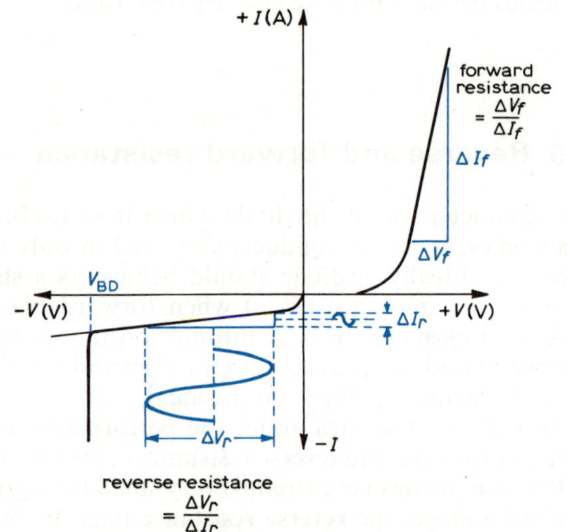

Fig. 2.5 Forward and reverse resistance

The reverse resistance of semiconductor diodes is normally many hundreds of kilohms.

The **forward resistance** of a diode is likewise determined by $\frac{\Delta V_f}{\Delta I_f}$ taken over the linear part of the forward characteristic. This value is very low, normally a few tens of ohms.

2.6 Peak inverse voltage (P.I.V.) and breakdown

There is a limit to the reverse voltage which can be applied to a diode. When this limit is reached there is a sharp increase in the reverse current (Fig. 2.5). The normally high reverse resistance suddenly becomes low; the **breakdown** of the junction has occurred. At breakdown there is a limit to the reverse voltage, V_{BD}.

Manufactures of semiconductor devices specify a safe maximum value to which the diode can be subjected without breakdown. This is called the **peak inverse voltage** (P.I.V.)

There are two main effects involved in the reverse breakdown of a *p-n* junction; zener breakdown and avalanche breakdown.

Zener breakdown occurs when the applied electric field is high enough to pull electrons from their covalent bond causing the production of electron-hole pairs in the depletion layer. Conduction is then possible because the depletion layer is no longer barren of charge carriers.

Zener breakdown occurs mainly at voltages below 6 V in *p-n* junctions that are heavily doped resulting in a very thin depletion layer. The exact value of the breakdown voltage is dependent mainly upon method of manufacture, type of material and surrounding temperature of the diode.

When the voltage is increased above the breakdown value, the diode may be permanently damaged if the maximum power rating is exceeded. Special diodes designed to withstand repeated reverse breakdown for the purpose of providing a constant voltage are called **zener diodes** (see Section 2.12).

Avalanche breakdown is probably the effect when breakdown first occurs for voltages above 6 volts. It generally affects semiconductors which are lightly doped and hence have a relatively wide depletion layer.

Minority electrons crossing the *p-n* junction are accelerated by the applied electric field. Often these electrons strike atoms in the crystal lattice with sufficient energy to dislodge valence electrons. The newly released electrons are accelerated by the field and become involved in other collisions, generating electron-hole pairs. More and more charge carriers are released in rapidly increasing numbers (the avalanche) and made available for conduction.

Avalanche breakdown also may be destructive if sufficient heat is generated.

2.7 Comparison of silicon and germanium diode characteristics

Fig. 2.4 shows the forward and reverse characteristics of typical silicon and germanium diodes. Particular differences to note are listed in Table 2.1.

Table 2.1 Comparison of diode values

	Si	Ge
Barrier potential, V_b	0.6 to 0.7 V	0.15 to 0.2 V
Reverse saturation current	μA	mA
Reverse resistance	very high	high

Review

Check your knowledge and understanding. You should:

a Be able to draw a *p-n* junction in the reverse and forward modes, indicating current flow in the diode and the external circuit.
b Appreciate the difference between the junction potentials of germanium and silicon diodes when connected in the forward bias mode.
c Appreciate and be able to draw the static characteristic for a diode.
d Appreciate the difference between the typical static characteristics for silicon and germanium diodes to illustrate the difference in forward voltage drop and reverse current.
e Understand the significance of the peak inverse voltage of a diode and the breakdown effect.

Exercise

2.1 To describe the formation of a *p-n* junction, arrange the following statements in a logical order (that is, which comes first, second, and so on),

a a depletion layer formed either side of the junction
b diffusion of majority carriers across the junction
c further diffusion prevented by electro-static repulsion of fixed charges in lattice
d recombination of carriers
e a *p-n* junction manufactured from a single piece of material.

2.2 Estimate to what value the potential barrier (V_b) becomes for both silicon and germanium when;

a a reverse bias voltage of 10 V is applied
b a forward bias of 0.5 V is applied

2.3 A silicon diode (V_b = 0.6 V) is rated at 1.8 W maximum power dissipation. (Calculate the maximum forward current accommodated by the diode.

2.4 Calculate the diode current if the diode in the diagram is

a silicon
b germanium

Hint: The current is limited by *R* when the diode is fully conducting and V_b is developed across D. Assume a suitable value for V_b.

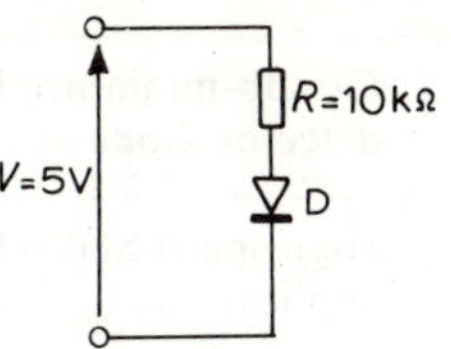

2.5 The following information refers to diode characteristics at 25 °C.

Forward

V(V)	0	0.5	0.6	0.65	0.7	0.75	0.8	0.85
I(mA)	0	0	0	1	3	10	40	100

The reverse characteristic is assumed to be a straight line between the points

V(V)	1	5
$I(\mu$A)	1	4

Determine the forward and reverse resistance.

2.6 If the reverse current of question 2.5 increases at 8% per degree C, sketch the new reverse characteristic (on same axes as before for comparison) for a temperature of 45 °C. Calculate the forward and reverse resistance at the higher temperature and compare with that at 25 °C.

2.8 Diode types

Many types of diodes are required in electronic circuits and in order to select the right diode for an application the following is specified by the manufacturer.

a maximum forward current rating
b peak inverse voltage
c forward voltage drop (V_b)
d switching speed (A measure of how quickly a diode can change from hard forward conduction to the negligible current of reverse bias. This speed is important for high frequency applications.)

In general it is possible to place diodes in two main categories of application, signal diodes and power diodes. **Signal diodes** are designed to operate efficiently for applied waveforms having high frequencies. Often they are used in communication circuits, before the signal has been amplified substantially, so they are required to handle only small voltages and

Table 2.2 Selection of diode types

Type No.	Description	Max. P.I.V.	Max. peak forward current	Typical reverse current at P.I.V. max.
OA90	Ge sub-miniature high frequency detector diode	30 V	45 mA	300 mA
BAW62	High speed Si diode for fast switching applications	75 V	225 mA	1 mA
OA210	General purpose Si rectifier diode	400 V	5 A	40 μA
BYX35	High voltage Si diode cased in a ceramic tube (oil-cooled). Designed for use in high voltage power supplies of X-ray, electron microscope and laser equipment.	25 kV	160 mA	—

currents. Hence their power ratings are low and their physical dimensions small (see Table 2.2).

A typical application of signal diodes is in a **demodulation** circuit. Here the diode is known as a **detector** because it separates the low frequency message signal (maybe voice or music) from the high transmission frequency or carrier wave.

Power diodes are designed to withstand high forward currents and have high P.I.V.s. They are extensively employed in rectifier circuits (see Section 2.10) which convert alternating current to direct current. For this reason, power diodes are generally used at low frequencies, particularly mains supply frequency at 50 Hz, and the switching speed is unimportant.

Some physically larger silicon diodes are designed to accommodate forward currents of up to several hundreds of amperes. Even though the forward resistance is low, the power dissipation ($I^2 R$) in the form of heat can be considerable. Because excessive temperature can seriously disrupt the crystal regularity of semiconductors, these diodes are often connected to a piece of metal called a heat sink (Fig. 2.6). The heat is then rapidly dissipated into the atmosphere, keeping the diode cool.

Silicon particularly is useful for power diodes as it can operate at higher temperatures than germanium (up to 150 °C for silicon, 70 °C for germanium). Silicon diodes are also available with P.I.V. ratings of up to thousands of volts. Germanium diodes are required sometimes because their small voltage drop across the potential barrier which is useful in particular applications.

Many other different types of semiconductor diodes are used for highly specialised applications such as computer circuitry or microwave communications. Descriptions of their operation lay outside the scope

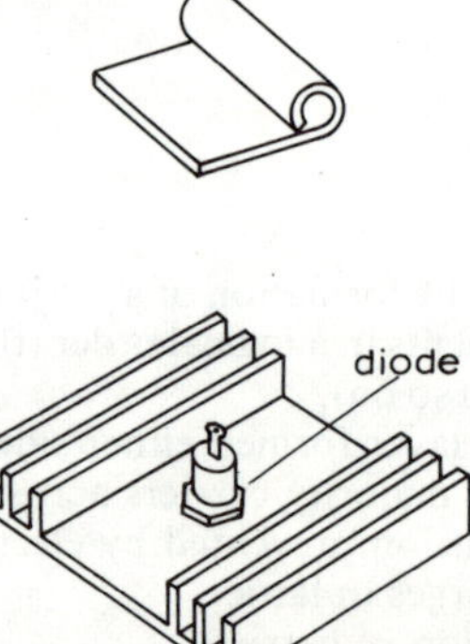

Fig. 2.6 Heat sinks

of this book, but names such as Schottky diodes, tunnel diodes, and varactor diodes, may become familiar to you.

2.9 Manufacture of the *p-n* junction diode

The first type of *p-n* junction diode was known popularly as a **cat's whisker**. The end of a fine strand of tungsten wire was touched on the surface of a piece of natural crystal called **galena**. By chance, this crystal exhibits the rectifying properties of a *p-n* junction. These crude diodes were used to detect signals in the early radio receivers at a time when little was understood about semiconductor theory. Usually the whisker had to be moved across the surface of the crystal to seek an effective *p-n* junction.

The cat's whisker led to the development of the more precisely manufactured **point-contact diode** (Fig. 2.7a). In this diode, the tungsten wire spring was retained, but the galena was replaced by a carefully processed *n*-type semi-conductor. Point contact

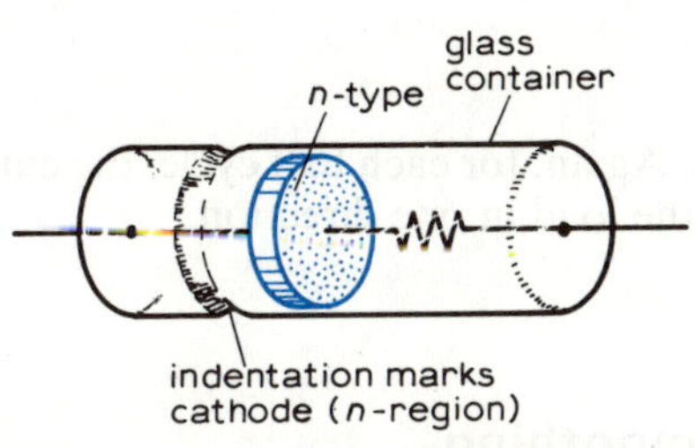

a Cat's whisker

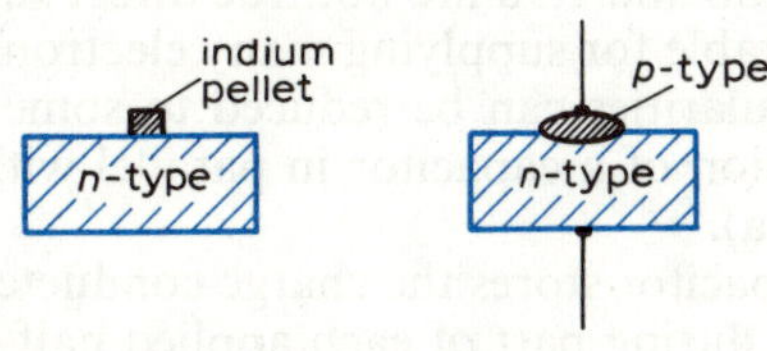

b Alloy junction

Fig. 2.7 Manufacture of diode

diodes respond to frequencies in the gigahertz region (greater than 10^9 Hz). Their current ratings are very low because of the small area of contact.

An important condition of an efficient *p-n* junction diode is that the crystalline structure across the junction is continuous with no disruptions between the bonds. This can be ensured by growing a crystal as described in Section 1.12. A suitable donor impurity is initially added to the melt to make the crystal *n*-type. After part of the crystal has been drawn, an excessive amount of an acceptor impurity is added to the melt which cancels the effect of the donor. The remainer of the grown crystal is then *p*-type.

A diode known as an **alloy junction** (Fig. 2.7b) is made by placing a small pellet of indium on an *n*-type silicon wafer. After being raised to a sufficiently high temperature, the molten indium mixes with a small part of the wafer making a region of alloy. When cooled, the alloy solidifies onto the remainder of the wafer, forming a *p-n* junction with a continuous crystalline structure.

Although junctions can be manufactured by a method called the planar technique, this is more commonly used for the more complex semiconductor devices. (This technology is discussed in relation to transistors and integrated circuits, see Chapters 3 and 15.)

Review

Check your knowledge and understanding. You should:

a Understand the terms used for the specification of diode types.
b Be able to state the simple applications and available range of signal and power diodes.
c Appreciate aspects of the methods of manufacture of *p-n* junction diodes.

2.10 Rectification

One very important use of diodes is in circuits which convert alternating voltage and current to direct voltage and current. The conversion of an alternating quantity to a direct quantity is done normally in two stages, called rectification and smoothing.

Figure 2.8 shows an alternating supply which provides an alternating current through the resistance R_L. The polarity of the alternating current reverses every half-cycle. The first step in converting an alternating current to a direct quantity is to ensure that the current flows through the circuit in only one direction. This is called **rectification**.

The simplest way of achieving this is by inserting

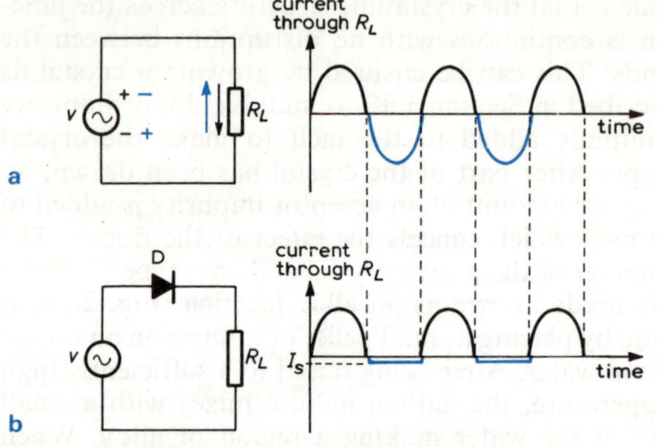

Fig. 2.8 Half-wave rectification
a Alternating current through R_L without diode
b Current through R_L with diode

a diode D between the supply and the load resistor as shown in Fig. 2.8b. During the positive half-cycle, the diode is forward biased (shown in black). As the polarity is reversed, the diode becomes reverse biased (shown in blue). Hence the current flows through the load in one direction only.

Using a diode as described is called **half-wave rectification**, as half the wave is lost. This, however, is not often used in practice. A more economical use is made of the full waveform by **full-wave rectification**.

A **full-wave rectifier bridge** circuit employing four diodes is shown in Fig. 2.9a. During each positive half cycle, the supply voltage is distributed across the diodes such that D2 and D3 conduct because they are forward biased; diodes D1 and D4 are reverse biased and can therefore be considered as open circuit because of their very high resistance. The path enabling current to flow through R_L is shown in Fig. 2.9b. When the supply polarity is reversed, diodes D1 and D4 conduct and diodes D2 and D3 can be considered open circuit (Fig. 2.9c).

Notice that in each half cycle, the current flows through R_L in the same direction. Each negative half cycle has been inverted and the combined waveform of current through R_L is shown in Fig. 2.9d.

A useful tip for memorising the bridge configuration is to remember that all the diode arrowheads point the same general direction (either right or left).

An alternative method of obtaining full-wave rectification using two diodes and a centre-tap transformer (one with an electrical connection at the centre of its secondary winding) is shown in Fig. 2.10. During the positive half-cycle (black), D1 conducts through R_L; when the polarity is reversed (blue), D2

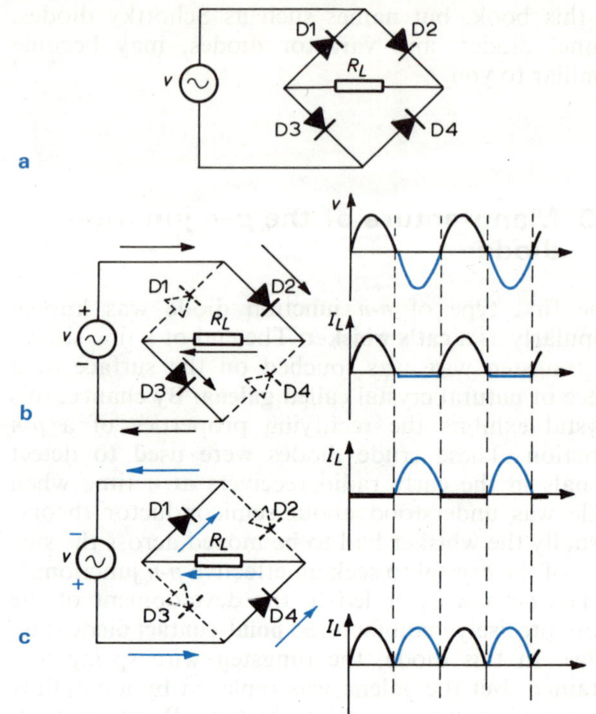

Fig. 2.9 Full-wave rectification

conducts. Again, for each half cycle, the current flows through the load in one direction.

2.11 Smoothing

The half-wave and full-wave rectified supplies shown in Figs. 2.8b and 2.9d are not true direct supplies and are unsuitable for supplying many electronic circuits. The irregularities can be reduced to some extent by the inclusion of a capacitor in parallel with the load (Fig. 2.11a).

The capacitor stores the charge conducted through the diode during part of each applied half cycle. The diode conducts only when the supply voltage v_s exceeds the voltage maintained by the capacitor v_R. In between peaks, when v_s is lower than v_R, the capacitor discharges its stored charge through the load, thus keeping the current more constant. The capacitor is sometimes referred to as a **reservoir capacitor**. It is said to **smooth** the waveform and also is called a **smoothing capacitor**.

The value of the capacitor is chosen such that the

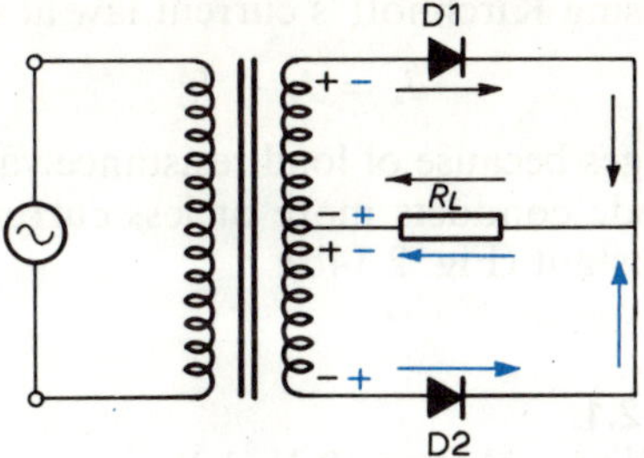

Fig. 2.10 Full-wave rectifier using centre-tap transformer

time taken for complete discharge (determined by the $C \times R_L$ time constant) is much longer than the periodic time T. This ensures that the current presented to the load falls only by a practicable minimum until the next charging pulse arrives.

Not all the alternating component can be removed from the waveform in this manner. The remaining unevenness of the output is called **ripple**. The full-wave rectifier leaves a much lower ripple than the half-wave because the capacitor is allowed to discharge for a much shorter time (Fig. 2.11b).

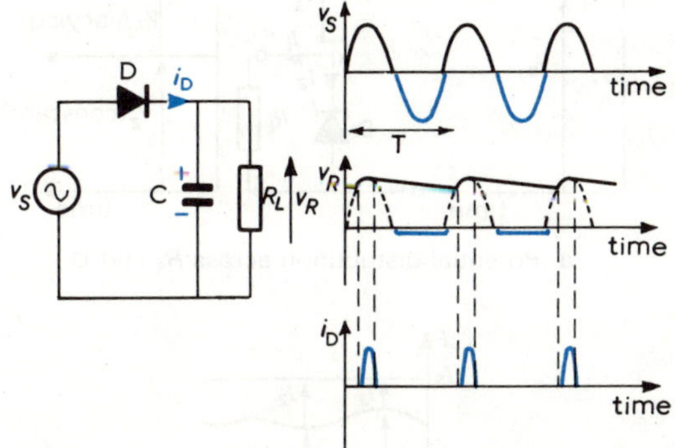

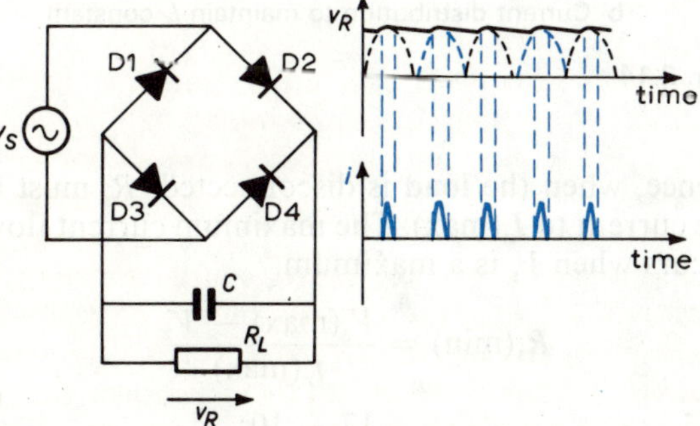

Fig. 2.11 Smoothing for half-wave and full-wave rectifier

The P.I.V. of the diodes used for rectification has to be a higher voltage than the peak-to-peak voltage of the supply. This is because when the supply voltage reaches a negative maximum, the reverse voltage can be almost the full peak-to-peak value because the voltage on the load side of the diode is held high by the capacitor.

The inclusion of inductors, capacitors and resistors in a smoothing network reduces the ripple more effectively than a single reservoir capacitor.

Review

Check your knowledge and understanding. You should:

a Be able to sketch the waveforms of applied a.c. voltage and load current for diode circuits which provide half-wave and full-wave rectification into a resistive load.

b Understand the effects of connecting a smoothing capacitor across the load resistor in half- and full-wave rectifier circuits upon the diode current waveform, the load current waveform, the load p.d. waveform and the inverse voltage applied to the diode.

2.12 Zener diode

Certain diodes are designed specifically to operate in the reverse breakdown region at precise voltages. These usually are called **Zener diodes**. (Even though at the higher breakdown voltages, avalanche breakdown may be the effect involved.) The breakdown characteristics of these diodes are more sharply defined than those of normal diodes (Fig. 2.12).

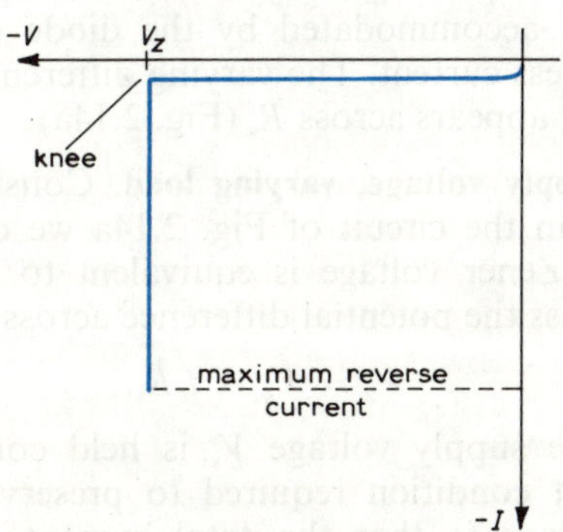

Fig. 2.12 Zener diode reverse characteristic

Zener diodes must be capable of supporting a substantial breakdown current before being damaged. The current that flows through the diode is limited by an external series resistor R_s as shown in Fig. 2.13. In this configuration, the reverse voltage developed across the diode does not significantly exceed the rated breakdown value, even if the applied voltage is increased. Therefore a constant voltage is available from a supply that may fluctuate.

The Zener diode is useful as a voltage stabiliser or as a voltage reference source in an electronic circuit. By careful manufacture, the breakdown can be fixed at any of a range of voltages between about 2 V and 200 V. To maintain a constant voltage, the diode is operated between the knee and the maximum current permitted by the power rating (Fig. 2.12). Below the knee, the diode would work as a normal reverse biased diode.

The Zener voltage V_z, which is applied across the load (represented by resistance R_L in Fig. 2.13) must be held constant under two conditions; varying supply voltage, fixed load, and fixed supply voltage, varying load.

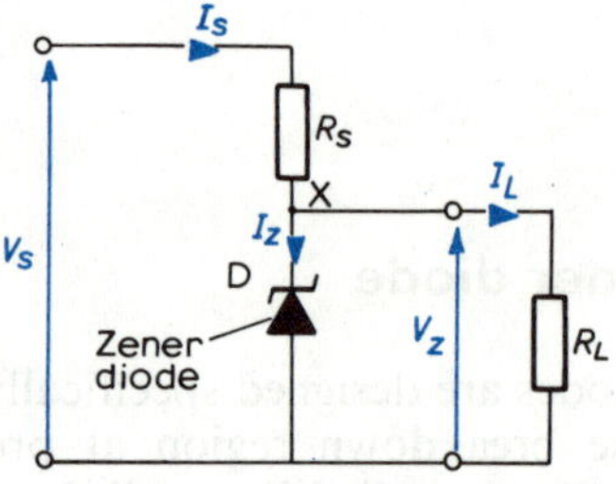

Fig. 2.13 The Zener diode voltage stabilising circuit

Varying supply voltage, fixed load. For V_z to remain unchanged, the load current I_L must be constant (from $V_z = I_L \times R_L$). Any variation in applied voltage V_s is accommodated by the diode conducting more or less current. The varying difference between V_s and V_z appears across R_s (Fig. 2.14a).

Fixed supply voltage, varying load. Considering the voltages in the circuit of Fig. 2.14a we can deduce that the Zener voltage is equivalent to the supply voltage less the potential difference across R_s

$$V_z = V_s - I_s R_s \tag{2.3}$$

When the supply voltage V_s is held constant, the important condition required to preserve constant load voltage is that the total current I_s remains unchanged for any variation of load resistance.

Also, using Kirchhoff's current law at node X,

$$I_s = I_z + I_L \tag{2.4}$$

If I_L changes because of load resistance variation, the Zener diode conducts more or less current to maintain I_s constant (Fig. 2.14b).

Example 2.1

A zener diode rated at 10 V, 1 W is used to regulate a variable supply voltage of 15 ± 2 V. Calculate the minimum series resistor required. Refer to Fig. 2.14.

Assume resistance of diode, when conducting, is negligible.

The power rating of the diode $P_z = I_z V_z$.

$$\text{Maximum permissible diode current } I_z = \frac{P_z}{V_z}$$

$$= 0.1 \text{ A.}$$

When the load is disconnected ($R_L = \infty$), all the supply current must be conducted by the diode.

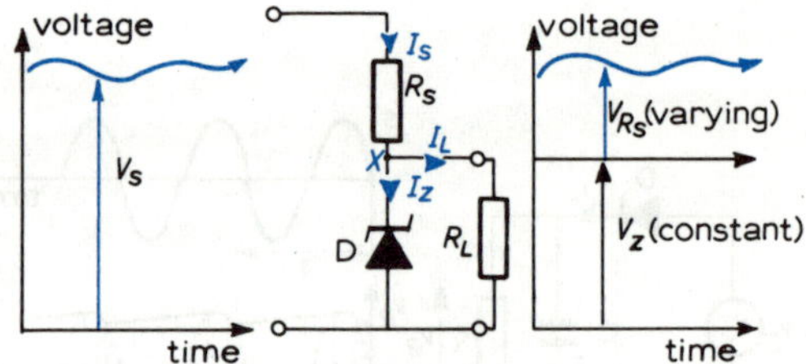

a Potential distribution across R_S and D

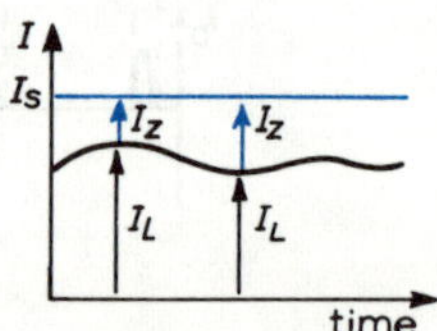

b Current distribution to maintain I_S constant

Fig. 2.14

Hence, when the load is disconnected, R_s must limit the current to I_z(max). The maximum current flowing occurs when V_s is a maximum.

$$R_s(\text{min}) = \frac{V_s(\text{max}) - V_z}{I_z(\text{max})} \tag{2.5}$$

$$= \frac{17 - 10}{0.1} = 70\ \Omega$$

The next highest preferred resistance value to 70 Ω (see Appendix 1) is 75 Ω. Assuming a tolerance of ±5%, the lowest value might be 75 − 3.75 = 71.25 Ω which is adequate for R_s.)

Example 2.2
Assuming that the supply is constant at 9.9 V calculate the range through which the load resistance may vary for a 6.8 V, 400 mW Zener diode to provide regulation.

As in Example 2.1, $I_z(\max) = \frac{P_z}{V_z} = \frac{400}{6.8}$

$$= 58.8 \text{ mA.}$$

$$R_s(\min) = \frac{V_s - V_z}{I_z(\max)}$$

$$= \frac{3.1}{58.8 \times 10^{-3}}$$

$$= 52.7\ \Omega$$

Next highest preferred value is 56 Ω

The minimum load resistor that may be connected is the value that takes most of the supply current, leaving the minimum current passing through the diode which prevents it falling below the knee. However, the knee current is so small that for our calculations it may be neglected.

Because I_z is assumed zero (which occurs for the condition of minimum load resistance), from Eqn. 2.4 we have

$$I_s = I_L \quad (2.6)$$

Now $$I_s = \frac{V_s - V_z}{R_s} \quad (2.7)$$

and $$I_L = \frac{V_z}{R_L(\min)} \quad (2.8)$$

Substituting Eqns. 2.7 and 2.8 into Eqn. 2.6,

$$\frac{V_s - V_z}{R_s} = \frac{V_z}{R_L(\min)} \quad (2.9)$$

$$\therefore R_L(\min) = \frac{R_s \times V_z}{(V_s - V_z)} = \frac{52.7 \times 6.8}{3.1}$$

$$= 115.6\ \Omega$$

$R_L(\max)$ is infinity because the series resistor R_s limits the maximum current through the Zener diode.

Review
Check your knowledge and understanding. You should:

a Be able to sketch a circuit diagram of a stabilised voltage source including a Zener diode and a series resistor.
b Know how to calculate the value of series resistor R in a simple Zener stabilising circuit for the conditions of varying supply voltage, fixed load and fixed supply voltage, varying load.

Exercise

2.7 A Zener diode 10 V, 400 mW is used to regulate a supply voltage of 12.5 ± 1.5 V. Calculate the minimum value of series resistor R_s.

2.8 Calculate the minimum power rating of a 15 V Zener diode used for regulating a supply variation of 20 ± 2.5 V when R_s = 91 Ω ± 5%.

Self assessment questions

The following questions are typical of those asked in an assessment test.

Multiple choice questions

2.9 Identify which of the following configurations forward biases the junction diode?

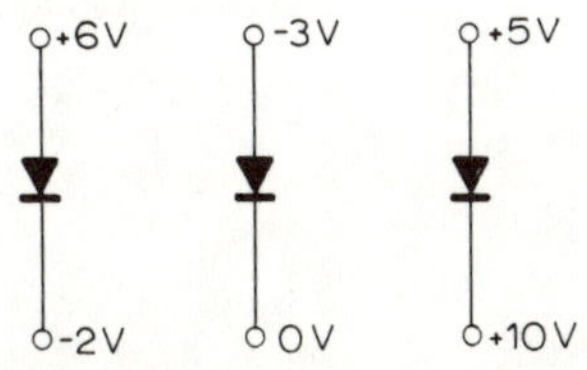

2.10 Identify typical barrier potentials for a *p-n* junction.
a 2 mV for germanium; 6 mV for silicon
b 0.2 V for germanium; 0.6 V for silicon
c 2 V for germanium; 6 V for silicon
d 200 mV for silicon; 600 mV for germanium

2.11 Identify which of the following names are used to describe the mechanisms of reverse breakdown;
a avalanche breakdown
b forward breakdown
c diffusion breakdown
d Zener breakdown
e abrupt breakdown

2.12 Identify which answer correctly completes the statement. A Zener diode is normally operated in
a the forward biased region below V_B
b the reverse breakdown region
c the reverse region before breakdown
d the forward region beyond V_B

2.13 Identify which answer correctly complete's the statement. Connecting a smoothing capacitor across the load resistance of a full-wave rectifier has the effect of
a increasing the output voltage
b reducing the ripple
c making the output half-wave rectified
d no noticeable effect at all

Short answer questions

2.14 Draw a *p-n* junction diode connected in
a the forward biased mode
b the reverse biased mode
Indicate for each mode which charge carriers contribute to the flow of current in the diode and the total current in the external circuit.

2.15 The following table gives the forward characteristic of a *p-n* junction diode. Plot the characteristic and determine the forward resistance of the diode.

V(V)	0	0.5	1	1.5	2.0
I(mA)	0	2.5	12.5	45	80

2.16 State simple application of
a signal diodes
b power diodes
c zener diodes

2.17 A Zener diode 6.8 V, 500 mW is used to regulate a supply voltage of 10 ± 1.5 V.
Calculate the minimum value of series resitor R_S.

2.18 If the supply voltage of question 2.17 is held constant at 10 V, calculate the permissible range of values of load resistance R_L.

3 The bipolar transistor

This chapter describes the bipolar transistor, its characteristics and the conditions necessary for its operations as an amplifier. The aims of this chapter are:

- to describe the bipolar transistor
- to provide an appreciation of the three circuit connections of a transistor and the characteristics they define
- to provide an understanding of the properties of a transistor
- to describe the construction of a transistor.

A graphical analysis is used to determine certain properties of the device which are used in the design of a simple practical electronic amplifier.

3.1 Introduction

An **amplifier** can be described as a device which provides a high energy output in relation to a small energy input. Mostly the output is expected to be proportional to input. The hi-fi amplifier, for example, is a commonly used piece of equipment which provides a substantial sound output in response to a very small input.

Of course amplifiers do not produce energy. The energy source for the output power is an external one such as a mains supply. It is useful to consider the amplifier as a *control* device; a small signal input controls an energy source such that it responds proportionately and provides a more powerful output.

Hi-fi amplifiers are pieces of complex electronic equipment and contain many electronic components. Among these components is the important semiconductor device called the transistor. This device in itself is an amplifier but needs to be connected to many other transistors and other components to provide the sophisticated level of control and amplification demanded of a modern hi-fi amplifier.

3.2 Physical description

Junction transistors are made from germanium or silicon and comprise a thin layer *p*-type semiconductor sandwiched between two pieces of *n*-type (called an ***npn*-transistor**), or a thin layer of *n*-type semiconductor in between two pieces of *p*-type (called a ***pnp*-transistor**).

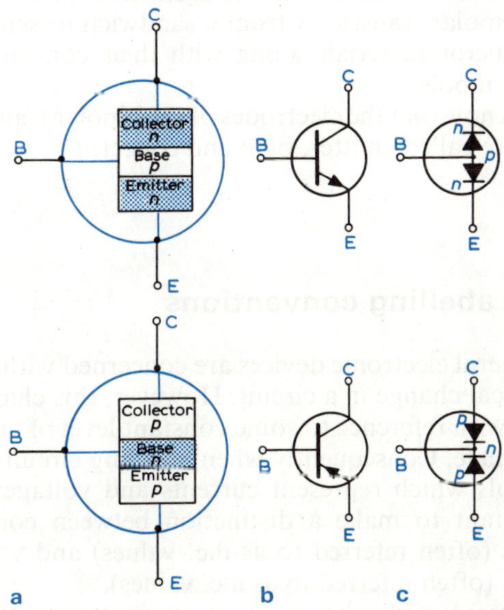

Fig. 3.1 **a** The semi conductor layers of *npn* and *pnp* transistors
b The circuit symbol of *npn* and *pnp* transistors
c Diode representation

The three semiconductor layers are identified by names which indicate their properties – **emitter**, **base** and **collector** – which will be explained later in this chapter. The connections to these three regions are labelled respectively E, B and C.

Fig. 3.1 shows the semiconductor layers of *npn* and *pnp* transistors, the **circuit symbols** and the representation of these transistors as diodes. The arrow on the symbols relates to the direction of current flow through the transistor.

The word **bipolar** means that two types of charge carrier are involved, holes and electrons, which are created in the *p*-type and *n*-type semiconductor materials (see Chapter 1). Although two types of charge carrier are required to form the *p-n* junctions, only one type of charge carrier is important for explaining the operation of the transistor; electrons for the *npn*, and holes for the *pnp*.

Although Fig. 3.1c represents a transistor as two diodes, it is not physically possible to synthesise an operational transistor in this way. Close contact between the regions is essential, also their thicknesses and doping densities are critical.

Review

Check your knowledge and understanding. You should:

- **a** Be able to sketch the arrangement of *npn* and *pnp* bipolar transistors from a sandwich of semiconductor materials along with their conventional symbols.
- **b** Know that the electrodes of the bipolar transistor are called emitter, base and collector.

3.3 Labelling conventions

In general electronic devices are concerned with some electrical change in a circuit. However, this change is often with reference to some constant level of current or voltage. Consequently when labelling circuits with symbols which represent currents and voltages it is important to make a distinction between constant values (often referred to as d.c. values) and varying values (often referred to as a.c. values).

Currents and voltages which are constant with time are represented by capital letters, e.g. I_E, I_C, V_{BE}, V_{CE}, and so on. The voltage subscripts indicate the two points between which the p.d. is defined. For example, V_{CE} is the p.d. between collector and emitter. The first subscript indicates the more positive potential with respect to the second subscript. Alternating or time varying values are represented by small letters, e.g. i_E, i_C, v_{CE}, and so on. These are used here to represent an instantaneous value caused by a varying electrical quantity (a signal).

3.4 Common-emitter mode

The purpose of the **common-emitter mode** of operation of a transistor is to provide conditions which allow a current to flow between the collector and emitter under control of a relatively small base current. It is called the common-emitter mode because the emitter terminal is made common to both input and output.

A circuit for considering the behaviour of transistor in this mode is shown in Fig. 3.2. The base-emitter junction is forward biased as shown in Fig. 3.2a. The arrow shows electrons injected into the emitter and passing into the base. The collector-base junction is simultaneously reverse biased (Fig. 3.2b). Although

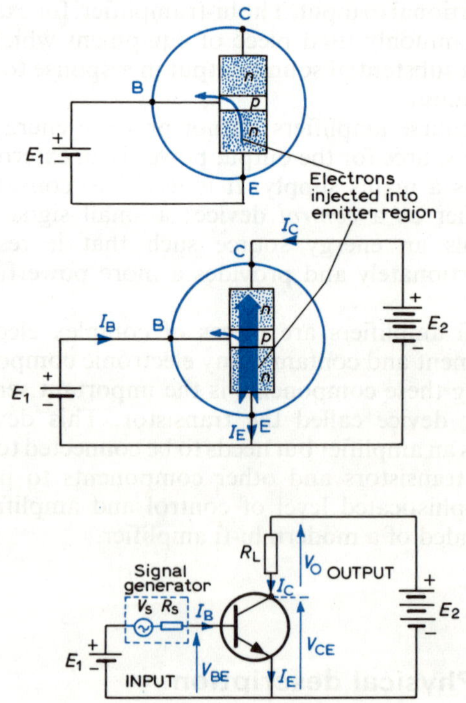

Fig. 3.2 The common-emitter mode of operation

E_2 is connected between the collector and emitter, the collector-base junction is reverse biased provided E_2 is greater than E_1. This results in a much greater number of electrons injected into the emitter.

Note that the two biasing conditions are essential for the operation of any bipolar transistor (*npn*, *pnp*, germanium or silicon).

The physical operation in this arrangement is as follows: electrons enter the emitter region from the negative side of E_1 and E_2. Electrons are majority carriers in the emitter and pass with ease through the forward biased emitter-base junction into the base region.

From here electrons can go one of two ways, out of the base terminal to the positive potential of E_1, or through the base-collector reverse biased region into the collector (electrons are minority carriers in the *p*-type base region and can therefore freely pass into the collector). They are then attracted by the positive potential of E_2.

Most of the electrons pass into the collector because the base region is very thin (typically 25 μm), and the doping density of the base is much lower than in the emitter and collector regions thus reducing the chance of electron-hole recombination.

The names of the regions now become apparent:

Emitter – *emits charge carriers into the base region*

Collector – *collects carriers from the base region*

The rate of charge movement determines the current magnitude. Generally, somewhere between 95% and 99.5% of the emitted electrons flow into the collector, the actual amount depends upon the design and manufacture of the device. The remainder of the electrons flow out of the base terminal.

Because three conductors meet at a point, we can deduce a relationship between the currents (Fig. 3.2b)

$$I_B + I_C = I_E \tag{3.1}$$

Note: The currents referred to here are conventional currents, the reverse direction of electron flow.

By varying the base potential with an input signal (Fig. 3.2c), the collector current is also varied. Fig. 3.3 shows a curve which represents the relationship between the base current I_B and the base-emitter voltage V_{BE}. You may recognise it as being similar to that of a forward biased *p-n* junction (see Section 2.4, Chapter 2).

The signal variations of I_B are followed by the much larger variations of I_C. If I_C is passed through a load resistor R_L an output voltage can be detected. Hence, the common-emitter connected transistor provides amplification of current and voltage.

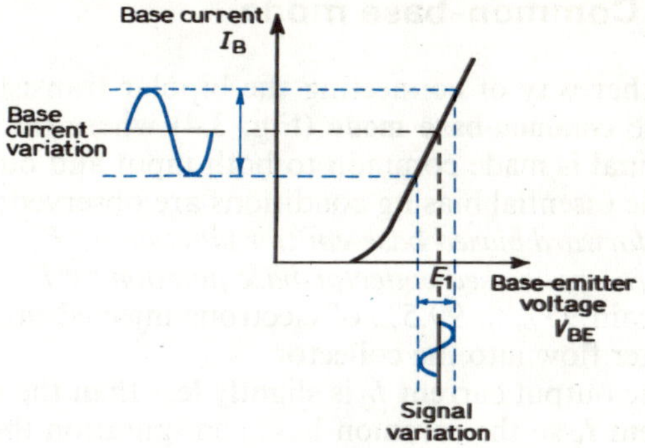

Fig. 3.3 Curve of I_B/V_{BE} (V_{CE} constant)

The **common-emitter current gain**

$$\beta = \frac{\text{a small change in } I_C}{\text{a small change in } I_B} = \frac{\Delta I_C}{\Delta I_B} \tag{3.2}$$

where the symbol Δ means a 'small change in'. β is a number and has no units.

Example 3.1

A transistor is connected in the common-emitter mode. If I_B varies by 1 μA, by how much does I_C vary when 98% of the emitter electrons reach the collector?

If the collector current is 98% of the total, the base current is 2% (from Eqn. 3.1). Therefore the collector current varies by an amount $\frac{98\%}{2\%} = 49$ times more than the base current.

From Eqn. 3.2

$$\Delta I_C = \beta \times \Delta I_B$$
$$= 49 \times 1\ \mu\text{A}$$
$$= 49\ \mu\text{A}$$

The two essential biasing conditions (forward biased base-emitter junction, reverse biased collector-base junction) are equally applicable to the *pnp* transistor. To comply with these operating conditions, the polarities of E_1 and E_2 are reversed. The above description of charge movement still applies for the *pnp* transistor, but consider holes injected into the emitter instead of electrons.

3.5 Common-base mode

Another way of connecting the bipolar transistor is in the **common-base mode** (Fig. 3.4) where the base terminal is made common to both input and output.

The essential biasing conditions are observed;

forward biased base-emitter junction by E_1
reverse biased collector-base junction by E_2.

Again 95% to 99.5% of electrons injected into the emitter flow into the collector.

The output current I_C is slightly less than the input current I_E so the common-base configuration there is no current gain. But because almost the same value of current is transferred from a region of low resistance R_{low} (forward biased base-emitted junction) to a region of high resistance R_{high} (reverse biased collector-base junction) a **power gain** is achieved (neglecting R_s and R_L in Fig. 3.4), i.e.

$$\text{input power } P_i = I_E{}^2 R_{low} \tag{3.3}$$

$$\text{output power } P_o = I_C{}^2 R_{high} \tag{3.4}$$

and as I_C is approximately equal to I_E the output power is greater than the input power

$$P_o > P_i \tag{3.5}$$

Hence the derivation of the name TRANSfer resISTOR.

If the emitter current is varied by a signal, the collector current likewise is varied. I_C is then passed through a load resistor R_L enabling an amplified voltage to be detected at the output. So, although the common-base mode of connection has a current gain less than unity, a voltage gain is achieved.

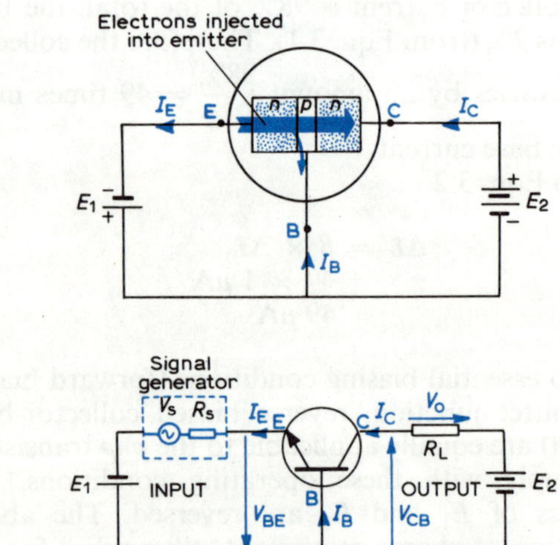

Fig. 3.4 The common-base mode of operation

The **common-base current gain**

$$\alpha = \frac{\text{a small change in } I_C}{\text{a small change in } I_E} = \frac{\Delta I_C}{\Delta I_E} \tag{3.6}$$

α is a number and has no units.

Example 3.2

The transistor used in Example 3.1 is connected in the common-base mode. If I_E varies by 50 μA, by how much does I_C vary when 98% of emitter electrons reach the collector?

$$\alpha = \frac{\Delta I_C}{\Delta I_E} = \frac{98\%}{100\%} = 0.98$$

$$\text{Therefore } \Delta I_C = \alpha \times \Delta I_E = 0.98 \times 50\ \mu\text{A} = 49\ \mu\text{A}.$$

Note: from Examples 3.1 and 3.2 that β and α are the respective current gains of the *same* transistor connected in different modes, the current conditions being the same in each.

Review

Check your knowledge and understanding. You should:

a Be able to sketch the circuit diagrams for the common-emitter and common-base modes of connection of a transistor.
b Appreciate the basic action of a transistor.
c Know the expressions for the current gains of a transistor connected in the common-emitter and common-base modes.

Exercise

3.1 Identify which answer correctly completes the following statement. The correct biasing conditions for a bipolar transistor are:

a C-B junction forward biased; B-E junction reverse biased
b C-E junction forward biased; B-E junction reverse biased
c B-E junction forward biased; C-B junction reverse biased
d different biasing conditions are required for each mode of connection

3.2 The change of collector current I_C of a bipolar transistor in the common-emitter mode is 0.6 mA for a change in base current I_B of 12.5 μA. Calculate the circuit current gain β.

a 12.5 **b** 48 **c** 21 **d** 210

3.3 The following refer to a transistor connected in the common-emitter mode:

a Base current = 2.6 μA; collector current = 0.12 mA. Calculate the current gain β.

b Collector current = 0.18 mA; current gain β = 85. Calculate the value of the base current.

c Emitter current = 1.98 mA; collector current = 1.95 mA. Calculate the current gain β.

3.6 Static characteristics

When a transistor is connected in a circuit there are several currents and voltages which provide information about its behaviour. In order to obtain comprehensive information about the characteristics of any *active* electronic device such as a transistor it is usual to use **static characteristics**. These are sets of graphs for the significant voltages and currents for the particular circuit configuration of interest. They are plotted in terms of three parameters. One is kept constant, the other two being varied over a practical range.

For example, in Fig. 3.5c a static characteristic of a common-emitter connected transistor has its base current I_B kept at a constant value and then the collector current I_C can be plotted for a range of values of collector-emitter voltage V_{CE}. This is then repeated for various other constant values of I_B. This set of static characteristics provides a comprehensive set of information about the transistor and the relationship between I_B, I_C, and V_{CE}.

Other sets of static characteristics can be plotted and although the information is about static conditions it can be deduced how the transistor behaves operationally in dynamic (or a.c.) conditions.

The following three sections provide information about the characteristics of transistors connected in the three modes.

3.7 Common-emitter static characteristics

Static characteristics are graphs which tell us how the transistor behaves when d.c. biasing voltages are applied. Initially the characteristics are considered when no a.c. input signal is present.

These characteristics are important for determining the operational properties of a transistor in a given set of circuit conditions. The main properties that can be determined from the graphs are voltage, current and power gains, and input and output resistances. Power dissipation within a transistor can also be calculated from the static characteristics.

The static characteristics for the common-emitter connection are described in relation to Fig. 3.5. V_{BE} and V_{CE} are the voltages appearing across the base-emitter and collector-emitter regions arising from the applied e.m.f.s E_1 and E_2 (as shown in Fig. 3.5a). I_E, I_B and I_C are the direct currents appearing in the emitter, base and collector respectively as a result of the applied voltages.

Three types of characteristics are shown for each transistor configuration. These are input characteristic, output characteristic, and transfer characteristic.

A circuit for obtaining the characteristics is shown

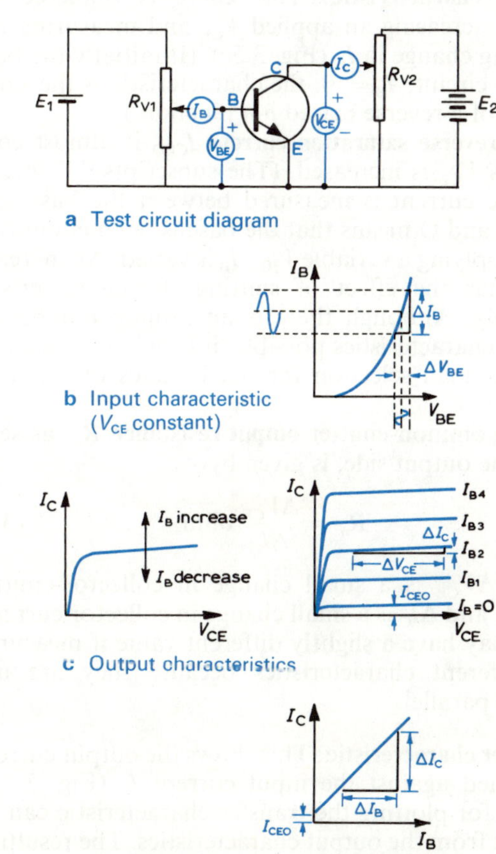

Fig. 3.5 Common-emitter static characteristics

in Fig. 3.5a. R_{V1} and R_{V2} are used to vary the potentials V_{BE} and V_{CE}. The values of E_1 and E_2 depend on the type of transistor (e.g. for the BC108 $E_1 = 2$ V and $E_2 = 20$ V).

Input characteristic: This is determined by measuring the base current I_B resulting from slowly increasing the base-emitter voltage V_{BE} (Fig. 3.5b). During these measurements, the collector is open circuit or the collector voltage held constant. (The resulting curve is the same as that of a forward biased *p-n* junction.)

A signal applied across the base-emitter junction would cause a variation of V_{BE} and I_B along the linear part of the input characteristic. The common-emitter **input resistance** R_i is therefore given by

$$R_i = \frac{\Delta V_{BE}}{\Delta I_B} \text{ ohm} \tag{3.7}$$

where ΔV_{BE} is a small change in base-emitter voltage and ΔI_B is a small change in base current.

Output characteristic: This curve is obtained by slowly increasing an applied V_{CE} and measuring the resulting change in I_C (Fig. 3.5c). (If initially the base is open circuit, $I_B = 0$, the characteristic is the same as that of a reverse biased *p-n* junction.)

The **reverse saturation current** I_{CEO} is almost constant as V_{CE} is increased. (The subscripts CE means that the current is measured between the base and emitter and O means that the base is open circuit.)

By applying a variable V_{BE}, I_B is varied. An increase in I_B has the effect of shifting the characteristic vertically. Although there is an infinite number of output characteristics possible, it is only necessary to show a small selection for fixed values of I_B. (Fig. 3.5c).

The common-emitter **output resistance** R_o, as seen from the output side, is given by

$$R_o = \frac{\Delta V_{CE}}{\Delta I_C} \text{ ohm} \tag{3.8}$$

where ΔV_{CE} is a small change in collector-emitter voltage and ΔI_C is a small change in collector current.

R_o may have a slightly different value if measured on different characteristics because they are not exactly parallel.

Transfer characteristic: This shows the output current I_C plotted against the input current I_B (Fig. 3.5d). Values for plotting the transfer characteristic can be derived from the output characteristics. The resulting plot may be slightly curved, but for our purposes may be assumed linear.

Consider the effect of a signal voltage v_s applied to the input. For a variation of base current ΔI_B, I_C can be seen to vary by an amount ΔI_C Fig. 3.5d. The signal current gain for the common-emitter mode is represented by the symbol β (Eqn. 3.2),

$$\text{where } \beta = \frac{\text{change in output current}}{\text{change in input current}}$$

$$= \frac{\text{change in collector current}}{\text{change in base current}}$$

$$= \frac{\Delta I_C}{\Delta I_B} \text{ for constant } V_{CE} \tag{3.9}$$

Example 3.3

If 98% of the emitter current reaches the collector, determine the forward current gain β. Assume a current variation of 1 mA occurs in the emitter.

$$\beta = \frac{\Delta I_C}{\Delta I_B}$$

$$= \frac{1 \text{ mA} \times 0.98}{1 \text{ mA} - 1 \text{ mA} \times 0.98}$$

$$\therefore \beta = \frac{0.98}{1 - 0.98}$$

$$= 49$$

3.8 Common-base static characteristics

Similar to the common-emitter mode, a set of characteristics can be drawn for transistors in the common-base mode. A circuit for measuring the characteristics is shown in Fig. 3.6a. The following measurements can be made from the curves of Fig. 3.6:

Input resistance for common-base,

$$R_i = \frac{\Delta V_{BE}}{\Delta I_E} \text{ ohm} \tag{3.10}$$

Output resistance for common-base,

$$R_o = \frac{\Delta V_{CB}}{\Delta I_C} \text{ ohm} \tag{3.11}$$

Current gain for common-base,

$$\alpha = \frac{\Delta I_C}{\Delta I_B}, \text{ for constant } V_{CB} \tag{3.12}$$

I_{CBO} is the collector-base reverse saturation current with zero emitter current.

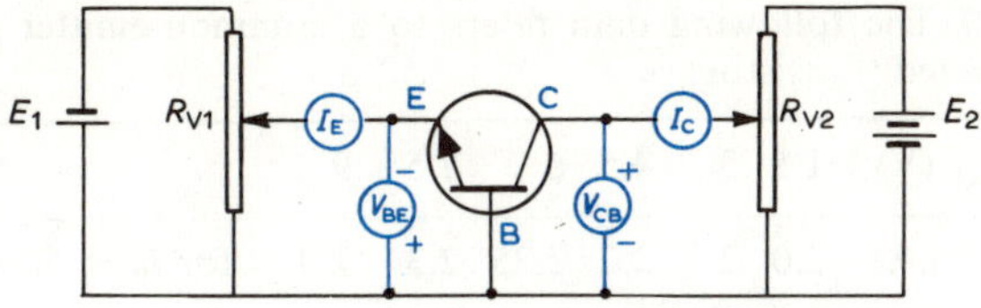

a Test circuit

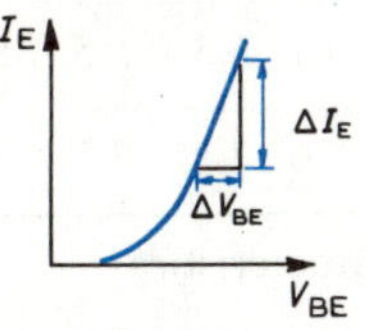

b Input characteristic (V_{CB} constant)

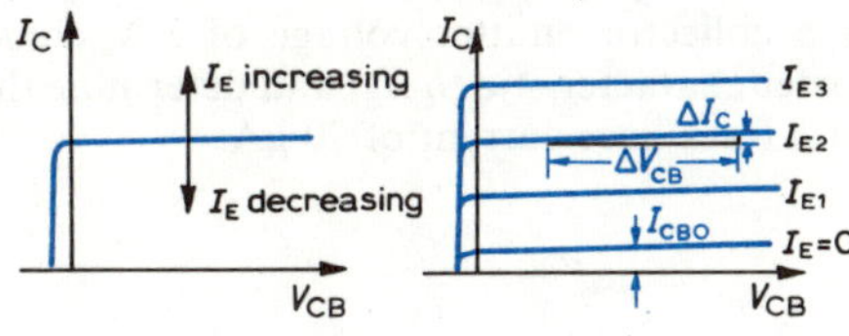

c Output characteristics

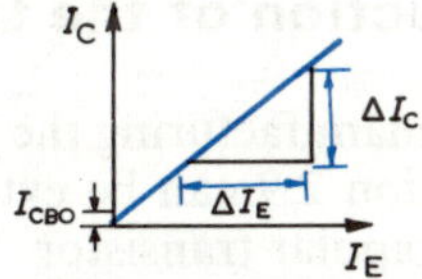

d Transfer characteristic (V_{CB} constant)

Fig. 3.6 Common-base static characteristics

Example 3.4

If 98% of the emitter current reaches the collector, calculate the forward current gain α. Assume a signal variation of 1 mA flows in the emitter.

$$\alpha = \frac{\Delta I_C}{\Delta I_E} = \frac{1 \text{ mA} \times 0.98}{1 \text{ mA}}$$

$$\therefore \alpha = 0.98$$

By comparing the solutions of Example 3.3 and 3.4 a relationship between α and β for the same transistor connected in the two modes can be obtained;

$$\beta = \frac{\alpha}{1 - \alpha} \tag{3.13}$$

By re-arranging Eqn. 3.13 for α

$$\alpha = \frac{\beta}{1 + \beta} \tag{3.14}$$

The derivation of these relationships is given in Appendix 2.

3.9 Common-collector mode

A third method of transistor operation which is not used as much as the others is the **common-collector** connection where the collector terminal is made common to both input and output.

In this mode the transistor exhibits a current gain but no voltage gain. It is useful for matching circuits of high resistance to circuits of low resistance with little loss of signal energy (e.g. high resistance microphone to the input of an amplifier). In these circumstances, the common collector is said to be working as a buffer amplifier (see Section 10.6, Chapter 10).

The common-collector mode sometimes is called an **emitter follower** because the emitter output closely follows the input variations at the base.

3.10 Comparison of input and output resistances

Table 3.1 shows the relative values of R_i and R_o for the common-emitter, common-base and common-collector modes of connection.

Table 3.1 Comparison of input and output resistances

Mode	Input resistance R_i	Output resistance R_o
Common-emitter	0.5 kΩ – 3 kΩ	10 kΩ – 50 kΩ
Common-base	50 Ω – 100 Ω	200 kΩ – 1 MΩ
Common-collector	50 kΩ – 1 MΩ	50 Ω – 100 Ω

Review

Check your knowledge and understanding. You should:

- **a** Appreciate the need for the static characteristics of transistors.
- **b** Be able to draw a common-emitter mode test circuit for determining static characteristics.
- **c** Know how to plot and understand the significance of the typical families of common-emitter characteristics:

  I_C/V_{CE} (output characteristic)
  I_C/I_B (transfer characteristic)
  I_B/V_{BE} (input characteristic)

d Be able to draw a common-base mode test circuit for determining static characteristics.
e Understand the method of obtaining the common-base mode static characteristics.
f Know how to plot and understand the significance of typical families of common-base characteristics:
I_C/V_{CB} (output characteristic)
I_C/I_E (transfer characteristic)
I_E/V_{BE} (input characteristic)
g Be able to determine the values of α and β from given characteristics.
h Know the relationship between α and β.
i Know how to obtain the values of output and input resistances from characteristics.
j Appreciate the relative values of input and output resistances for the three modes of connection.

Exercise

3.4 Calculate the respective common-emitter current gains β corresponding to the common-base gains α of:
a 0.98 **b** 0.965 **c** 0.99 **d** 0.975

3.5 Identify which answer correctly completes the statement. The input characteristic of a bipolar transistor connected in the common-base mode shows:
a how I_E varies with change in V_{BE} (V_{CB} constant)
b how I_E varies with change in V_{CB} (V_{CE} constant)
c how I_E varies with change in I_B (V_{CB} constant)
d how I_E varies with change in I_C (V_{CE} constant)

3.6 A transistor is connected as shown in Fig. 3.8.

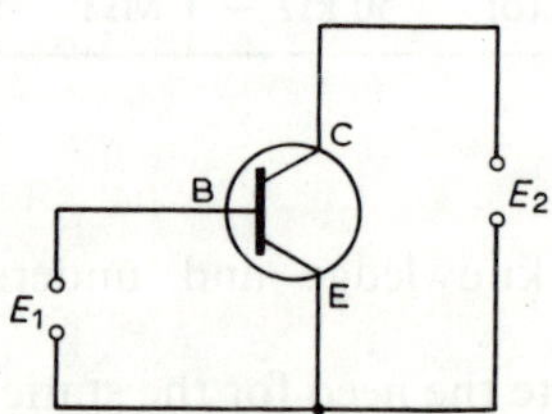

Fig. 3.8

a Label the polarities of E_1 and E_2 to bias the transistor correctly if it is **i** *npn* **ii** *pnp*.
b Which direction is the arrowhead to represent **i** *npn* **ii** *pnp*?
c When E_2 is disconnected, sketch the graph of I_B/E_1. What is this graph called?

3.7 The following data refers to a common-emitter connected transistor:

V_{CE} (V)	1.5	3	4.5	6	7.5	9	
I_C (mA)	2.0	2.1	2.2	2.25	2.3	2.4	for $I_B = 50\ \mu A$
I_C (mA)	3.0	3.1	3.2	3.35	3.45	3.5	for $I_B = 72.5\ \mu A$
I_C (mA)	4.7	4.9	5.1	5.3	5.5	5.7	for $I_B = 112\ \mu A$
I_C (mA)	6.4	6.7	7.0	7.25	7.5	7.8	for $I_B = 151\ \mu A$

Plot the output characteristics.

a Calculate the output resistance for the characteristic curve when $I_B = 72.5\ \mu A$.
b For a collector emitter voltage of 3 V construct the transfer characteristic (I_C/I_B) and determine the current gain β for a base current of 70 μA.

3.11 Construction of the transistor

The methods of manufacturing the *p-n* junction diode described in section 2.9 can be extended to the construction of the bipolar transistor.

The earliest form of transistor consisted of two wires in contact with a piece of semiconductor. These transistors are used rarely these days. They are called **point contact transistors** (Fig. 3.7a) and have two tungsten wires touching the *n*-type semiconductor base, making the three regions of emitter, base and collector.

Junction transistors are produced by **alloying** indium pellets (see Section 2.9) to both sides of a semiconductor wafer (Fig. 3.7b) to form the semiconductor sandwich. The depth of alloy penetration into the wafer is carefully controlled so that a very thin base region remains. A bipolar transistor can also be *grown* from the melt. The three regions are created by adding appropriate impurities to the melt to make the drawn crystal alternately *p*-type and *n*-type.

Today, the most important and commonly used method of manufacturing transistors is by **planar technology**, meaning that the devices are constructed in layers or planes. The steps involved with this technology are as follows (with reference to Fig. 3.7c).

A wafer of *n*-type silicon is first placed in an atmosphere of oxygen and the temperature increased until a thin skin of insulating oxide (silicon dioxide) is formed on the exposed surface.

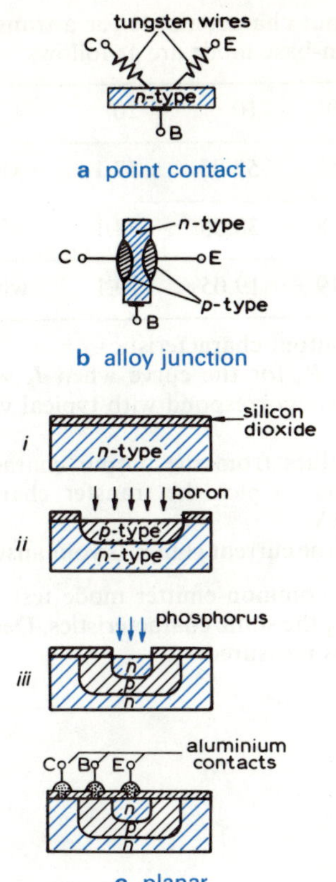

Fig. 3.7 Various methods of construction of bipolar transistor

A small opening called a **window** is made in the oxide layer by chemical etching. The wafer is then exposed to a vapour containing an acceptor impurity (e.g. boron) at a high temperature. Impurity atoms are absorbed through the window into the silicon by a mechanism known as diffusion, forming a *p*-type layer.

The wafer is oxidised again and an even smaller window etched above the *p*-region. Another diffusion is made with a vapour containing an *n*-type impurity (e.g. phosphorus), resulting in another *n*-region. Aluminium contacts are then deposited for the connection of leads to the emitter, base and collector regions.

Planar technology is ideally suited for automated manufacture because the processing steps are done on the wafer from only one side. Many wafers can be simultaneously oxidised, etched and diffused. The process can be extended to make multi-component **integrated circuits** (see Chapter 15). Planar transistors have the advantage of having a large collector region and are therefore better able to dissipate heat.

Self assessment questions

The following questions are typical of those asked in an assessment test.

Multiple choice questions

3.8 Identify the relationship between α and β:

a $\alpha = \dfrac{\beta}{1+\beta}$

b $\alpha = \dfrac{\beta}{1-\beta}$

c $\alpha = \dfrac{1}{1+\beta}$

d $\alpha = \dfrac{1}{\beta - 1}$

3.9 Identify which answer correctly completes the statement. The output characteristic of a bipolar transistor connected in the common-emitter mode is a graph of:

a I_C against a base of I_B for a fixed value of V_{CE}
b I_C against a base of V_{CE} for a fixed value of I_B
c V_{CE} against a base of I_E for a fixed value of I_B
d I_B against a base of V_{CE} for a fixed value of I_C

Short answer questions

3.10 The following refer to a transistor connected in the common-base mode:

a An emitter current of 0.5 mA produces a collector current of 0.49 mA. Calculate the current gain α and the value of the base current.
b If the current gain α is 0.97 and the collector current 0.75 mA, calculate the value of emitter current.
c If a current gain of 0.98 was obtained when the emitter current was 670 μA, calculate the collector current.
d For a current gain of 0.97 and an emitter current of 2.5 mA calculate the collector current and the base current.

3.11 The following data for a common-emitter connected transistor was obtained with a constant collector current.

I_B (μA)	0	0	2	14	45	155
V_{BE} (V)	0	0.2	0.4	0.6	0.8	1.0

Plot the input characteristic curve and calculate the input resistance for a base-emitter voltage of 0.8 V

3.12 The output characteristics of a transistor connected in the common-emitter mode are linear between the points in the table:

V_{CE} (V)	2	25	
I_C (mA)	4.5	5.5	when $I_B = 50\ \mu A$
I_C (mA)	14	15.5	when $I_B = 150\ \mu A$
I_C (mA)	23.5	26	when $I_B = 250\ \mu A$

a Plot the output characteristics.
b Calculate the output resistance R_o for the curve when $I_B = 150\ \mu A$. Does your answer correspond with typical values quoted in Table 3.1?
c Derive values from the output characteristics which enable you to plot the transfer characteristic when $V_{CE} = 15$ V.
d Determine from the transfer characteristic the current gain β. Is your answer typical of β?

3.13 The output characteristics for a transistor connected in the common-base mode are as follows:

V_{CB} (V)	0	10	20	
I_C (mA)	57	57.05	57.1	when $I_E = 60$ mA
I_C (mA)	38	38.05	38.1	when $I_E = 40$ mA
I_C (mA)	19	19.05	19.1	when $I_E = 20$ mA

a Plot the output characteristics.
b Calculate R_o for the curve when $I_E = 40$ mA. Does your answer correspond with typical values quoted in Table 3.1?
c Derive values from the output characteristics which enable you to plot the transfer characteristic when $V_{CB} = 10$ V.
d Calculate the current gain α. Is your answer typical of α?

3.14 Sketch a common-emitter mode test circuit diagram for determining the static characteristics. Describe how each characteristic is measured.

4 The small-signal transistor amplifier

This chapter introduces simple amplifiers which are able to respond to small input signals. The aims of this chapter are:

- to describe a simple single-stage transistor amplifier and its function
- to provide an appreciation of the amplification of signals
- to provide an understanding of the calculations necessary to obtain the various parameters of a transistor amplifier and the values of the circuit components
- to appreciate the limitations of single-stage small-signal transistor amplifiers.

4.1 Introduction

In the previous chapter the bipolar transistor characteristics were considered and the concept of current gains introduced. In order that the transistor can be used in practice as an amplifier it is necessary to incorporate it in a circuit with various **passive** components, such as resistors and capacitors and the appropriate power supplies.

Simple amplifiers which are able to respond to **small input signals** are called **small-signal amplifiers** and make use of only small responses from the transistor. This means that an undistorted (or linear) output can be obtained from the transistor. It is arranged that the output response of the transistor, in relation to its characteristics, ranges only over a small suitable linear portion.

These techniques are essential in many applications (such as audio amplification) where the output must be as faithful a reproduction of the input controlling signal as possible.

The appreciation and understanding of the small signal amplifier relies heavily on a good knowledge and understanding of the static characteristics described in chapter three.

4.2 Basic common-emitter amplifier: no input signal

The **transistor common-emitter amplifier** is used often because of its high current and voltage gains. A practical circuit is shown in Fig. 4.1. It is called a **single-stage** amplifier because the amplification process takes place through one *active* device (transistor) only.

To describe how it functions, first consider only the direct currents and voltages present in the circuit when no input signal is applied. These are known as **quiescent values** (meaning 'standing values') and their symbols bear the subscript Q (e.g. quiescent base current I_{BQ} and quiescent collector voltage V_{CQ}).

To avoid having two batteries E_1 and E_2 as in Fig. 3.2 (Section 3.4), the biasing potentials can be fixed with one d.c. supply $+V_{CC}$ and a base biasing resistor R_B.

The supply voltage is distributed across R_B and the base-emitter junction such that the potential appearing across the base has the same effect as E_1 in Fig. 3.2. That is, it forward biases the junction.

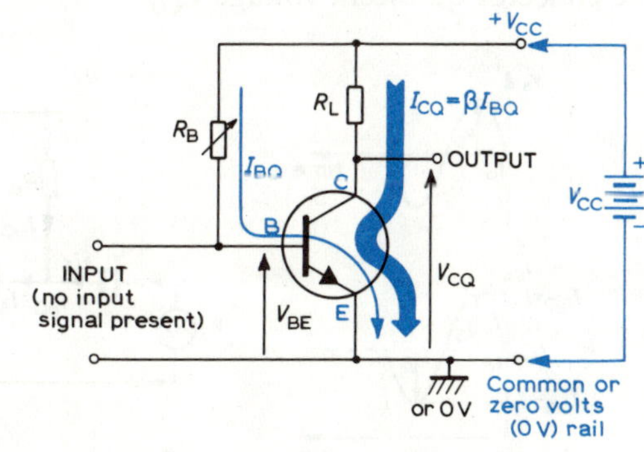

Fig. 4.1 The basic common-emitter amplifier stage

R_B is used to set I_{BQ} which flows from the supply and through the transistor. Because the base-emitter *p-n* junction is forward biased, a barrier voltage V_{BE} appears across it.

Therefore, by Ohm's Law, the current in the base is determined by the p.d. across R_B divided by the value of R_B.

$$I_{BQ} = \frac{V_{CC} - V_{BE}}{R_B} \quad (4.1)$$

R_L is the collector load resistance through which flows the quiescent collector current I_{CQ}. This current is equal approximately to the base current amplified by the current gain β (neglecting the very small reverse saturation current I_{CEO}).

$$I_{CQ} \approx \beta\, I_{BQ} \quad (4.2)$$

To correctly bias the *npn* transistor, a positive d.c. supply voltage is used in conjunction with R_B and R_L (Fig. 4.1). (For a *pnp* transistor amplifier a negative supply voltage $-V_{CC}$ is required.)

The output is taken from the collector. The direct voltage standing there when no signal is present is V_{CQ} (see Fig. 4.1). On the output side of the amplifier V_{CC} gives rise to two potentials, V_{CQ} and the potential across R_L

$$\text{i.e.} \quad V_{CC} = V_{CQ} + I_{CQ} R_L \quad (4.3)$$

Re-arranging for V_{CQ}

$$V_{CQ} = V_{CC} - I_{CQ} R_L \quad (4.4)$$

Example 4.1

A common emitter amplifier has a supply voltage $V_{CC} = +6\text{ V}$, $R_L = 1\text{ k}\Omega$, and $R_B = 720\text{ k}\Omega$. Calculate

a the base quiescent current I_{BQ}
b the collector quiescent current I_{CQ}
c the collector quiescent voltage V_{CQ}

The transistor is a BC 108 with a β of 334. Assume $V_{BE} = 0.6$ V.

a From Eqn. 4.1

$$I_{BQ} = \frac{V_{CC} - V_{BE}}{R_B} = \frac{6 - 0.6}{720 \times 10^3} = 7.5 \times 10^{-6}\text{ A.}$$

b From Eqn. 4.2

$$I_{CQ} \approx \beta I_{BQ} = 334 \times 7.5 \times 10^{-6} = 2.5 \times 10^{-3}\text{ A} \quad \text{or} \quad 2.5\text{ mA}$$

c From Eqn. 4.4

$$V_{CQ} = V_{CC} - I_{CQ}R_L = 6 - 2.5 \times 10^{-3} \times 10^3 = 3.5\text{ V}$$

Note: These are quiescent values and ideally should not vary. A signal has not yet been introduced.

4.3 Signal amplification by the common-emitter amplifier

Having set up the quiescent values in a common-emitter amplifier a small input signal value v_s is applied at the base (Fig. 4.2a). (For the purposes of this text it is convenient to consider a sinusoidal signal of peak value V_s and to neglect the small internal resistance of the signal source.)

The capacitor C_i is included at the input to block the d.c. from the supply being drained away through the circuitry which provides the signal. This would upset the biasing conditions of the transistor. The value of C_i is chosen so that it has a low reactance

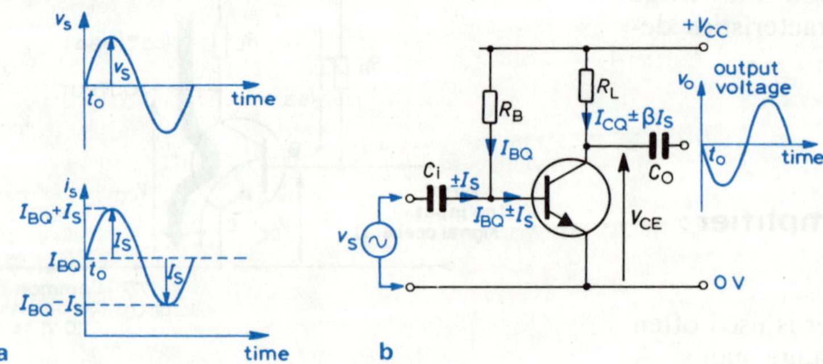

Fig. 4.2 The introduction of a signal V_s to the common-emitter amplifier

over the signal frequency range, allowing the signal to pass with little reduction in amplitude.

The signal voltage v_s causes a sinusoidal current to flow in the base (Fig. 4.2a). This causes a variation about the quiescent base current. The signal is amplified and appears in the collector causing a variation about the quiescent collector current. This can be shown as follows.

The signal current adds to or subtracts from I_{BQ} up to a peak value of I_S assuming that R_B is so high that its effect on the signal can be neglected. The total base current varies between $I_{BQ} + I_S$ and $I_{BQ} - I_S$, that is, $I_{BQ} \pm I_S$.

Thus the total collector current

$$I_C \approx \beta(I_{BQ} \pm I_S) \approx \beta I_{BQ} \pm \beta I_S \quad (4.5)$$

Because $I_{CQ} \approx \beta I_{BQ}$ (Eqn. 4.2) the total collector current

$$I_C \approx I_{CQ} \pm \beta I_S \quad (4.6)$$

Therefore it can be seen that the input signal current $\pm I_S$ appears in the collector circuit amplified by the amplification factor β.

As the amplified signal current passes through R_L, the p.d. developed across it varies in a similar manner, giving a sinusoidal output voltage at the collector. The collector-emitter voltage V_{CE} and collector current I_C change from the quiescent values V_{CQ} and I_{CQ} owing to the variations brought about by the introduction of a signal.

Therefore Eqn. 4.4 becomes

$$V_{CE} = V_{CC} - I_C R_L \quad (4.7)$$

where V_{CE} and I_C are changing values or variables.

When the signal voltage is rising, the base current increases followed by an increase in the collector current. This means that the potential across R_L becomes larger and from Eqn. 4.7 it follows that the output voltage V_{CE} becomes smaller. For a falling signal voltage, V_{CE} would increase. The output voltage wave form is inverted or 180° out of phase with the input (as shown in Fig. 4.2). As the message information of say speech or music is carried by the amplitude and frequency of the signal, this phase difference often is unimportant.

Review

Check your knowledge and understanding. You should:

a Be able to draw the circuit diagram of a single stage amplifier having a load resistor R_L and know that its supply voltage $V_{CC} = I_C R_L + V_{CE}$.

b Understand that bias is required in a small signal amplifier to give a selected quiescent operating point on the output characteristics.

c Be able to sketch the circuit diagram and explain the action of a simple bias arrangement consisting of a resistor connected between V_{CC} and base.

d Appreciate the effect of a small sinusoidal current input on the quiescent conditions of a single-stage amplifier.

e Know that voltage phase inversion occurs between the input and output signal of a single-stage amplifier.

Exercise

4.1 Identify which of the following statements are true and which are false:

a Resistor R_B and supply V_{CC} determine the quiescent base current.

b Resistor R_L has no useful purpose in the common-emitter amplifier.

c The common-emitter amplifier provides a voltage gain but no current gain.

d An input signal causes a variation about the quiescent conditions.

e Phase inversion occurs between the input and output signals of a common-emitter amplifier.

4.2 Explain why it is necessary for the supply voltage of a *pnp* transistor amplifier to be made negative.

4.4 Load line (d.c.)

Figure 4.3 shows the output characteristics of a transistor (BC 108) and *not* the characteristic of the single-stage amplifier containing the transistor.

The characteristics of the amplifier depend upon the values of the passive circuit components and the supply value (V_{CC}) as well as the properties of the transistor.

There is a technique by which all the possible combinations of I_C, V_{CE} and I_B for a particular set of circuit components can be obtained from the output characteristics of the transistor. This involves drawing a line on the characteristics defined by R_L and the quiescent point Q defined by R_B and V_{CC}. The line, (called a **load line**) and its construction and application are explained in this section.

Consider what happens to the position of the quiescent point (point Q Fig. 4.3) as the quiescent base current is slowly varied by adjusting a variable R_B (also see Fig. 4.1).

When R_B is so high that it can be considered an

open circuit, I_{BQ} is zero. Because there is no current in the base to be amplified, I_{CQ} is also zero.

Eqn. 4.4 $\quad V_{CQ} = V_{CC} - I_{CQ}R_L$
becomes $\quad V_{CQ} = V_{CC} \qquad (4.8)$

Under these conditions the quiescent point has moved to the lowest possible point on the horizontal axis of the output characteristics, marked as Q_1 on Fig. 4.3a.

When R_B is reduced to a very low value, I_{BQ} increases which increases I_{CQ}. The collector conduction becomes so high that a point is reached where the transistor itself can be regarded as a short circuit (zero resistance). When this happens, V_{CQ} is zero and all of the supply V_{CC} is developed across R_L. The collector current I_{CQ} is then limited only by R_L.

Hence $$I_{CQ} = \frac{V_{CC}}{R_L} \qquad (4.9)$$

(This can be verified by substituting $V_{CQ} = 0$ in Eqn. 4.4.)

The quiescent point has moved to its highest possible theoretical position, marked as Q_2 on Fig. 4.3a.

As R_B is varied between the values of zero and infinite resistance, the quiescent point traces a straight line between the extreme points of Q_1 and Q_2. This is the **d.c. load line**, so called because no a.c. signal has been considered.

The load line can be verified mathematically by rearranging Eqn. 4.4 for I_{CQ} as follows

$$I_{CQ} = -\frac{1}{R_L} V_{CQ} + \frac{V_{CC}}{R_L} \qquad (4.10)$$

Eqn. 4.10 is a straight line equation following the mathematical relationship $y = mx + c$ where the constants are $m = -\frac{1}{R_L}$ (negative slope) and $c = \frac{V_{CC}}{R_L}$. By, in turn, inserting the conditions of $V_{CQ} = 0$ and $I_{CQ} = 0$ Eqns. 4.8 and 4.9 are obtained which define the two end points of the line.

Under normal working conditions, when a signal is amplified, the quiescent point is not expected to move. It lies on the load line at a point fixed by V_{CC} and R_L (Eqn. 4.10). The load line gives the permissible variations of I_C and V_{CE} about the point Q (Fig. 4.3b).

When a signal is applied the quiescent values are no longer being considered and Eqn. 4.10 becomes

$$I_C = \frac{-1}{R_L} V_{CE} + \frac{V_{CC}}{R_L} \qquad (4.11)$$

where I_C and V_{CE} are varied about I_{CQ} and V_{CQ} by the signal.

This is still a straight line equation, so the variations brought about by the signal are along the load line.

From the signal variations shown in Fig. 4.3b a graphical determination of the amplifier current gain, voltage gain and hence power gain can be obtained.

$$\textbf{Current Gain } A_i = \frac{\text{change in output current}}{\text{change in input current}}$$

$$= \frac{\Delta I_C}{\Delta I_B} \qquad (4.12)$$

where ΔI_B is the peak-to-peak signal current in the base giving rise to ΔI_C, the peak-to-peak signal current in the collector.

$$\textbf{Voltage gain } A_v = \frac{\text{change in output voltage}}{\text{change in input voltage}}$$

$$= \frac{\Delta V_{CE}}{\Delta V_S} \qquad (4.13)$$

where ΔV_S is assumed to be the peak-to-peak signal voltage at the base and ΔV_{CE} is the peak-to-peak signal variation at the collector. If r.m.s. values are required,

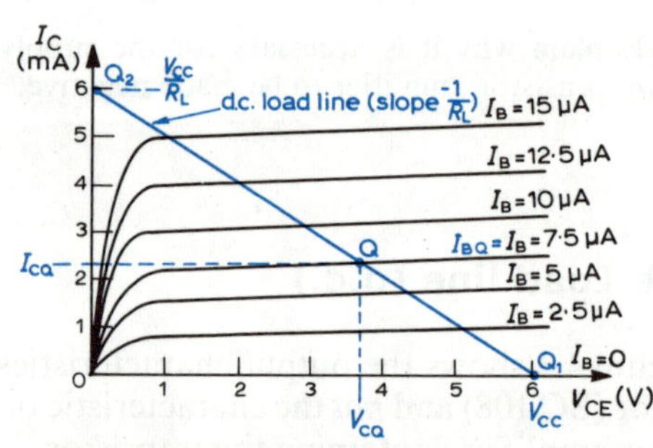

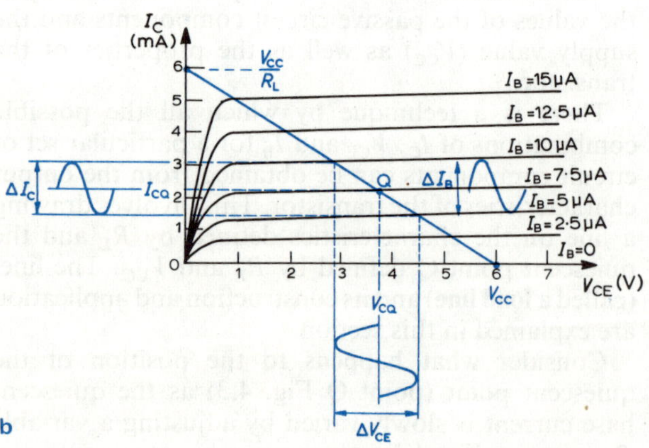

Fig. 4.3 The d.c. load line on the output characteristics

the peak-to-peak variations are multiplied by the factor $\dfrac{1}{2\sqrt{2}}$

$$\text{e.g. } I_C(\text{r.m.s.}) = \frac{\Delta I_C}{2\sqrt{2}}$$

$$\textbf{Power gain } A_p = \frac{\text{output power}}{\text{input power}}$$

$$= \frac{\text{r.m.s. output signal voltage} \times \text{r.m.s. output signal current}}{\text{r.m.s. input signal voltage} \times \text{r.m.s. input signal current}}$$

$$= \frac{V_{CE}(\text{r.m.s.}) \times I_C(\text{r.m.s.})}{V_S(\text{r.m.s.}) \times I_B(\text{r.m.s.})} \qquad (4.14)$$

Because the r.m.s. and peak-to-peak values are all related by the same constant $(1/2\sqrt{2})$, Eqn. 4.14 becomes:

$$A_p = \frac{\Delta V_{CE}/2\sqrt{2} \times \Delta I_C/2\sqrt{2}}{\Delta V_S/2\sqrt{2} \times \Delta I_B/2\sqrt{2}} = \frac{\Delta V_{CE} \times \Delta I_C}{\Delta V_S \times \Delta I_B} \qquad (4.15)$$

Close examination of Eqn. 4.15 reveals the power gain to be equal to the voltage gain multiplied by the current gain (Eqns. 4.12 and 4.13).

$$A_p = A_v \times A_i \qquad (4.16)$$

Power gains are often quoted in the logarithmic units of decibels (dB)

$$\text{where} \quad A_p = 10 \log_{10}\left(\frac{\text{output power}}{\text{input power}}\right)\text{dB} \qquad (4.17)$$

4.5 True loads and the a.c. load line

The analysis of Section 4.4 is adequate if R_L is the only resistance loading the transistor. This would be the case if the amplifier stage were driving a loud-speaker and R_L was representative of its resistance.

Often the output signal must be fed to a further stage of amplification or to some other form of load, in which case the output would be taken from between the collector and emitter (Fig. 4.4a). The true load would then be represented by R_{LL}. The inclusion of capacitor C_o prevents a d.c. drain from the supply V_{CC} through the load which would change the quiescent conditions. The value of C_o is chosen to have a small reactance over the applied frequency range, causing minimal reduction of the output signal amplitude.

Because of the presence of C_o, the d.c. conditions are the same as discussed in Section 4.4. The purpose of R_L is now to fix the d.c. conditions in the collector. So the d.c. load line plotted on the output characteristics remains unchanged and the quiescent point Q is found in the same place. However, the signal variation about Q no longer follows the d.c. load line. A new line called the **a.c. load line** or **dynamic load line** must be plotted which includes the effect of R_{LL} (Fig. 4.4c).

To an a.c. signal, the supply of V_{CC} appears as a large capacitance with a very low reactance, virtually a short-circuit at the frequencies used. Here is an example of where a.c. 'sees' a circuit entirely differently from d.c. Therefore the signal encounters R_L and R_{LL} in parallel (Fig. 4.4b). Whereas the d.c. load line had a slope of $-1/R_L$, the a.c. load line has a slope of $-1/R_p$

$$\text{where} \qquad R_p = \frac{R_L \times R_{LL}}{R_L + R_{LL}} \qquad (4.18)$$

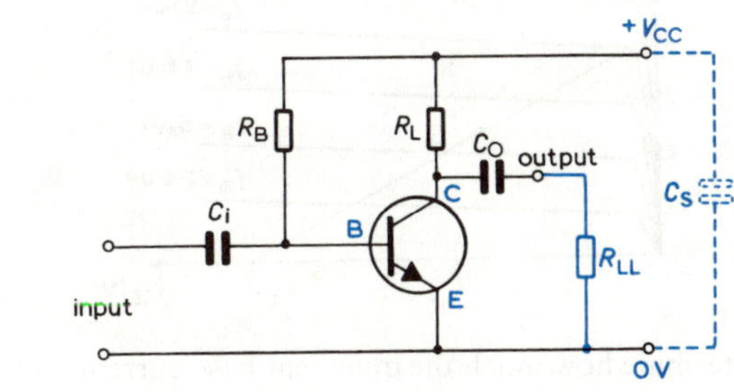

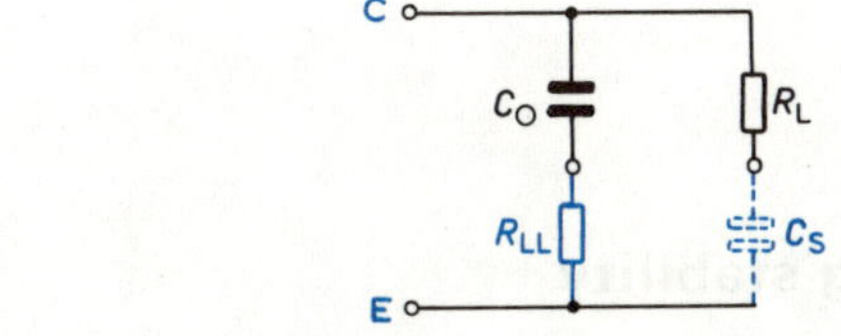

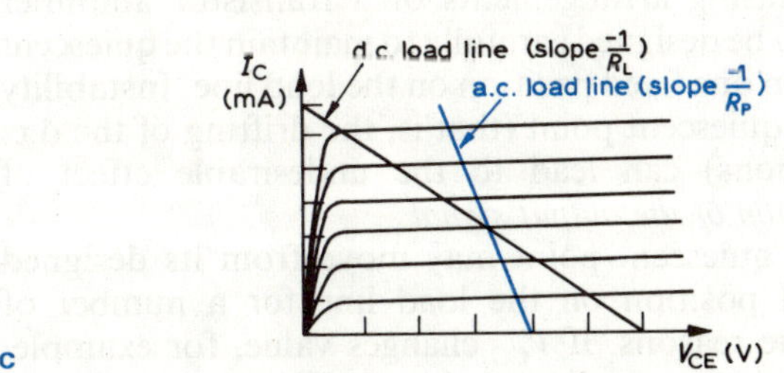

Fig. 4.4 A true load and the a.c. load line

Review

Check your knowledge and understanding. You should:

a Be able to construct the load line on a given set of output characteristics of a common-emitter amplifier for a stated value of load resistance.

b Be able to estimate the r.m.s. voltage output from the load line for given quiescent conditions and input signals.

c Know how to calculate the voltage gain A_v from the static characteristics and load line.

d Know how to calculate the current gain A_i from the static characteristics and load line.

e Know how to calculate the power gain A_p in dB.

Exercise

4.3 For a given transistor and supply voltage state which sole factor determines the slope of the load line.

4.4 Using information from the figure calculate R_B for a silicon *npn* transistor (Assume $V_{BE} = 0.6$ V).

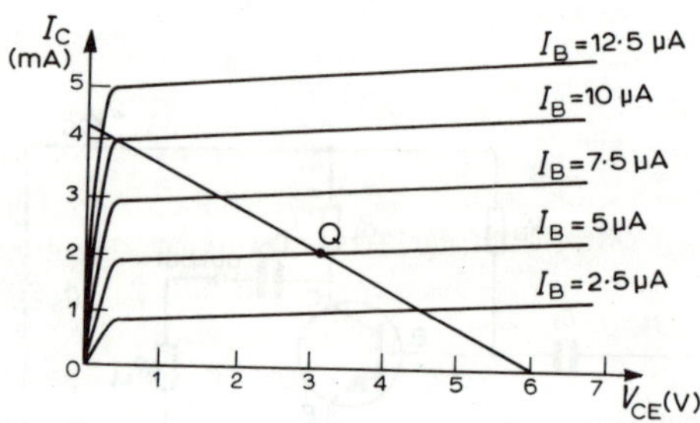

4.5 Determine how much the quiescent base current would have to increase before a 5 μA peak-to-peak base signal caused a distorted output.

4.6 Biasing stability

The biasing arrangements of a transistor amplifier need to be designed carefully to maintain the quiescent point in one fixed position on the load line. Instability of the quiescent point (that is, the drifting of the d.c. conditions) can lead to the undesirable effect of *distortion of the output signal.*

The quiescent point may move from its designed central position on the load line for a number of possible reasons. If V_{CC} changes value, for example, the quiescent conditions change. This may cause the transistor to operate in the non-linear region of its output characteristics (Fig. 4.5), leading to distortion or even clipping of the output signal.

Clipping can occur at the top of the waveform or at the bottom. The maximum collector current is limited by V_{CC}/R_L, even if the base signal tries to push it higher (Fig. 4.5a). This is known as **saturation** of the transistor. Because I_C cannot go negative, the lower part of the waveform may be clipped. This is sometimes referred to as bottoming the transistor.

In extreme cases instability of the quiescent point occurs with **thermal runaway**. This is caused by the dissipation of power, which occurs mainly in the collector, raising the temperature of the transistor. As the temperature rises, electron-hole pairs are generated increasing the reverse saturation current I_{CBO}, which is added to I_B and I_C. More power is dissipated, raising the temperature even further, creating more charge carriers, and so on. The quiescent point therefore creeps up the load line. This effect continues to snowball. If left unchecked, the maximum power rating is exceeded, causing permanent damage to the device. Heat sinks are therefore connected to susceptible transistors. These rapidly dissipate the heat, helping prevent thermal runaway.

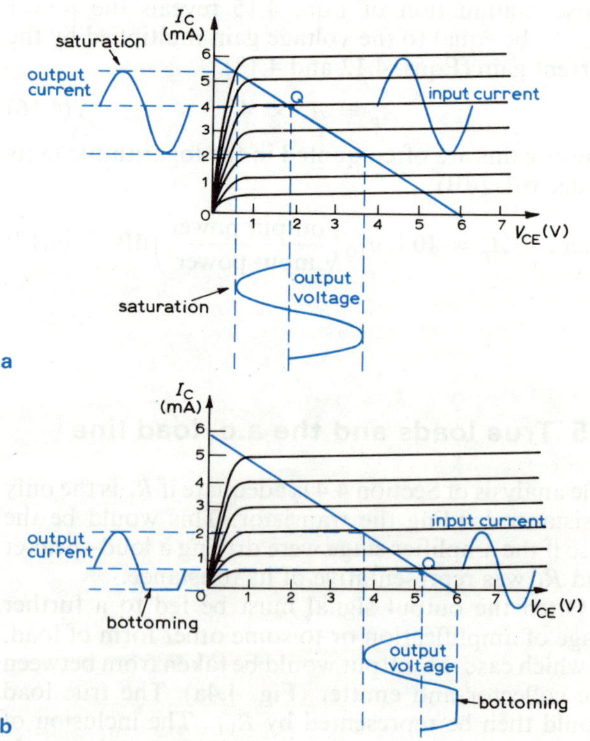

Fig. 4.5 Distortions of wave form by extreme position of quiescent point

4.7 Self or automatic biasing

The base resistor for bias described in this chapter (Fig. 4.1), offers little protection against thermal run-away. The stability is poor, consequently it is more usual to use other methods. Two common forms of **self biasing** or **automatic biasing** are **collector-base resistor biasing**, and **base voltage-divider biasing**.

Collector-base resistor biasing: The base bias resistor is connected to the collector rather than the supply rail (Fig. 4.6a), the value of I_B depends on V_{CE},

$$\text{i.e.} \quad I_B = \frac{V_{CE}}{R_B} \qquad (4.19)$$

If I_C increases because of temperature, the potential difference across R_L becomes larger, reducing V_{CE} (remember $V_{CE} = V_{CC} - I_C R_L$) and hence lowering I_B. A reduction in I_B tends to keep the amplified current I_C at its original value and better stability is achieved.

A certain reduction of amplification is inevitable with this type of biasing as true signals also are suppressed.

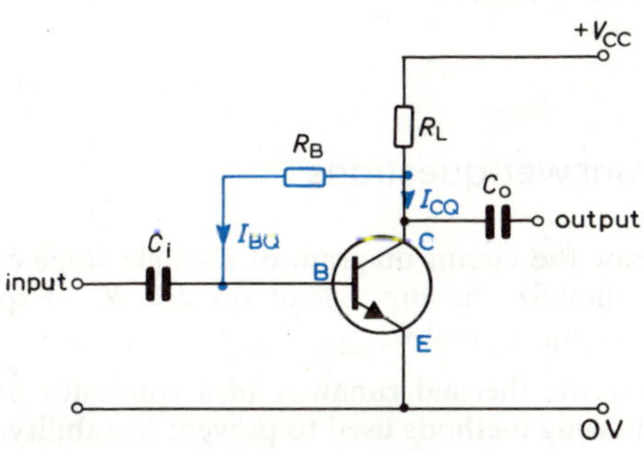

a Collector-base resistor biasing

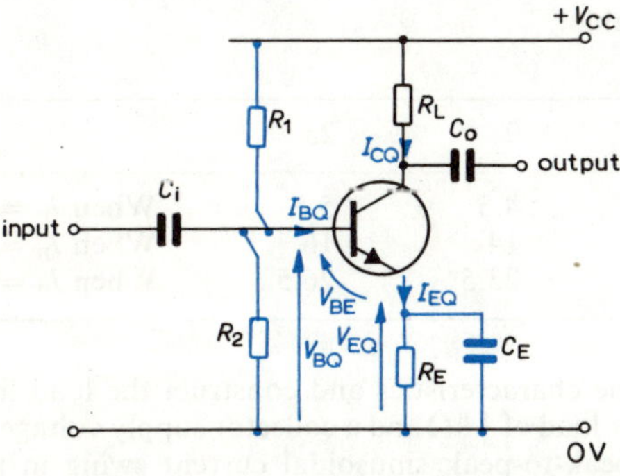

b Base voltage-divider biasing

Fig. 4.6 Alternative biasing methods of transistor

Base voltage-divider biasing: The base voltage-divider biasing network (Fig. 4.6b) provides the best stability of the methods discussed here. Resistors R_1 and R_2 function as a potential divider. That is, they share V_{CC} between them in proportions depending on their respective values. The quiescent voltage standing at the base V_{BQ} is held constant by this arrangement.

We can deduce from Fig. 4.6b that

$$V_{BQ} = V_{BE} + V_{EQ} \qquad (4.20)$$

where V_{BE} is the base-emitter voltage and V_{EQ} is the potential difference across resistor R_E.

The action of stabilising is as follows:

1 If I_{CQ} increases because of temperature, I_{EQ} also increases (because $I_{EQ} \approx I_{CQ}$), therefore V_{EQ} is larger.

2 From Eqn. 4.20 we see that V_{BE} would then decrease because V_{BQ} is held constant.

3 As I_{BQ} depends on V_{BE} (see Fig. 3.3, the input characteristic), the base current is reduced.

4 If I_{BQ} is lowered, it follows that I_{CQ} must be reduced, tending to compensate for any drift in the quiescent conditions and prevent the distortion and thermal runaway discussed earlier.

Note that the sequence of events 1 to 4 begins with the increase of collector current which eventually initiates its own control of stability.

The suppression of a true signal by this method is avoided by the inclusion of the decoupling capacitor C_E. The emitter capacitor ensures that the a.c. signal by-passes R_E. Because the capacitor is a block to d.c. the stabilising mechanism functions for d.c. only and the signal is unaffected.

If the d.c. load line of this arrangement was now plotted on the output characteristics, the slope would be $-1/(R_L + R_E)$.

Review

Check your knowledge and understanding. You should:

a Understand how the output of a transistor amplifier becomes distorted owing to the instability of the quiescent point.

b Appreciate the effect of thermal runaway of a transistor.

c Understand why heat sinks are used for transistors.

d Know the advantages and operation of collector-base resistor and base voltage-divider biasing of single-stage transistor amplifiers.

Exercise

4.6 The purpose of the emitter by-pass capacitor in a common-emitter amplifier is to do which of the following?

a increase the effective value of the emitter resistance at low frequencies
b prevent distortion
c to increase the effective frequency range of the amplifier
d to prevent signal dissipation in the emitter circuit

4.7 The output characteristics of a transistor connected in the common-emitter mode are assumed linear between the points given in the following table. These are to be plotted and used for questions 4.7 to 4.12

V_{CE}(V)	1.5	9.0	
I_C(mA)	1.0	1.25	when $I_B = 25\ \mu A$
I_C(mA)	2.0	2.4	when $I_B = 50\ \mu A$
I_C(mA)	3.0	3.75	when $I_B = 75\ \mu A$
I_C(mA)	4.2	5.25	when $I_B = 100\ \mu A$
I_C(mA)	5.3	6.5	when $I_B = 125\ \mu A$
I_C(mA)	6.4	7.8	when $I_B = 150\ \mu A$

The transistor is used in a basic amplifier circuit as shown in Fig. 4.1. For a supply voltage $V_{CC} = +8$ V plot on the same set of characteristics, load lines for

a $R_L = 1\ k\Omega$
b $R_L = 1.33\ k\Omega$
c $R_L = 2\ k\Omega$

4.8 On each load line of question 4.7 mark on quiescent points when $R_B = 98.66\ k\Omega$ and $V_{BE} = 0.6$ V. From the characteristics, determine the values of V_{CQ} and I_{CQ} for each quiescent point.

4.9 Determine which of the quiescent points determined in question 4.8 would cause a distorted output waveform for an input current swing (I_B) of 50 μA peak-to-peak.

4.10 Identify which load line of question 4.7 would permit maximum current swing without distortion. Determine the peak-to-peak base current which could be accommodated by this load line.

4.11 Calculate the current gains, voltage gains, and power gains (all as ratios) for each of the load lines in question 4.7 when a base voltage swing of 50 mV peak-to-peak gives rise to a base current swing of 50 μA peak-to-peak.

4.12 Using the load line plotted for $V_{CC} = 8$ V and $R_L = 2\ k\Omega$ in question 4.7 mark on the quiescent points for $R_B = 49.33\ k\Omega$, 59.2 kΩ, 74 kΩ, 98.66 kΩ, 148 kΩ, and 290 kΩ.
Determine which of the quiescent points would allow a base current swing of 50 μA peak-to-peak without causing distortion of the output signal.

Self assessment questions

The following questions are typical of those asked in an assessment test.

Multiple choice questions

4.13 Identify which of the following voltage equations is relevant to a common-emitter transistor amplifier:

a $V_{CC} = I_C R_L + V_{CE}$
b $V_{CE} = I_C R_L + V_{CC}$
c $V_{CC} = I_C R_L + V_{CB}$
d $V_{CB} = I_E R_L + V_{CC}$

4.14 Identify which of the following gives the two end points of a load line:

a V_{CC}/R_L and V_{CE}
b V_{CE}/R_L and V_{CC}
c V_{CC}/R_L and V_{CC}
d V_{CE}/R_L and V_{CE}

4.15 Identify which two factors determine the quiescent base current:

a V_{CC} and R_B
b V_{CC} and R_L
c V_{CE} and R_B
d V_{CE} and R_L

4.16 Identify which answer completes the statement. The effect of temperature increasing the collector current to the point of damaging the transistor is known as:

a temperature instability
b output distortion
c thermal drift
d thermal runaway.

Short answer questions

4.17 Draw the circuit diagram of a single stage common-emitter amplifier having a load resistor R_L. Explain the purpose of R_B, C_i and C_o.

4.18 Describe thermal runaway in a transistor amplifier. Sketch biasing methods used to prevent instability.

4.19 The output characteristics of a transistor connected in the common-emitter mode are linear between the points in the table:

V_{CE}(V)	2	25	
I_C(mA)	4.5	5.5	When $I_B = 50\ \mu A$
I_C(mA)	14	16	When $I_B = 150\ \mu A$
I_C(mA)	23.5	26.5	When $I_B = 250\ \mu A$

Draw the characteristics and construct the load line for a collector load of 1 kΩ and a collector supply voltage of 30 V. If the peak-to-peak sinusoidal current swing in the base circuit is 200 μA about a mean value of 150 μA and the input voltage swing is 800 mV peak to peak, determine current gain A_i, voltage gain A_v and power gain A_p.

5 Thermionic valves

This chapter describes two thermionic devices, the thermionic diode and triode. The aims of this chapter are as follows:

- to describe the phenomenon of thermionic emission
- to describe the construction of the thermionic diode
- to provide an appreciation of the characteristics of the thermionic diode
- to provide a comparison of thermionic and semiconductor devices
- to describe the construction of the thermionic triode
- to provide an appreciation of the characteristics of the thermionic triode
- to provide an understanding of the amplifying properties of the triode valve.

5.1 Introduction

Thermionic diodes and **triodes** have similar properties to the semiconductor diode and the transistor. They were developed long before the semiconductor devices commonly used today (although semiconductor diodes such as the point contact diode were the very first electronic devices employed for their unidirectional properties).

Thermionic valves are not used frequently today as they have been largely superseded by semiconductor devices. However, certain applications of **thermionic emission**, on which thermionic valves depend, are in general use for devices such as the cathode-ray tube.

5.2 Thermionic emission

Thermionic emission describes a phenomenon where electrons are emitted from a substance because of its high temperature (Fig. 5.1).

Many electrons move around freely within certain conducting materials. On the application of heat electron velocities can increase until those with sufficient energy escape from the surface of the material. If the material is in a vacuum or surrounded by very low pressure gas these electrons form a cloud around the surface of the material. The cloud is called a **space charge**.

Even though electrons are continually entering and leaving the space charge, the total number of electrons within it remain constant for the same temperature. As the temperature of the material is increased, more electrons are given to the space charge. The formation of a space charge is more pronounced in a vacuum where electrons do not collide with air molecules.

Tungsten, thoriated tungsten and oxide coated surfaces are used commonly as emissive materials.

Tungsten is a commonly used emissive material. Although it requires a high working temperature of approximately 2200 °C, it is suitable for high voltage thermionic devices (operated over 10 kV).

Thoriated tungsten is tungsten coated with thorium oxide. This material exhibits more thermionic emission than tungsten alone and at a lower temperature of 1500 °C. Thoriated tungsten is used in thermionic devices for voltages of less than 10 kV.

Oxide coated emitting surfaces are made from a mixture of barium and strontium oxides deposited on

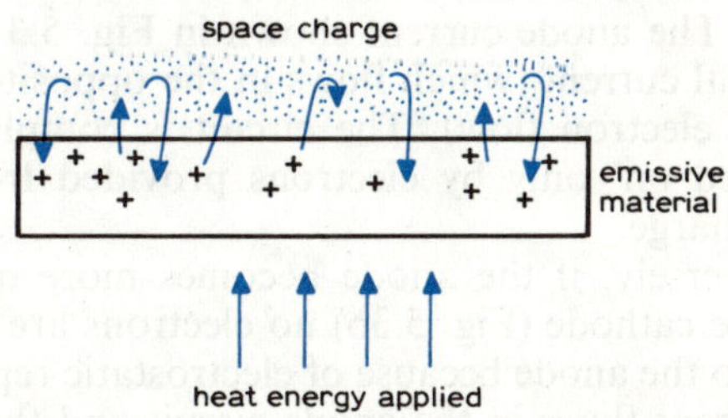

Fig. 5.1 Thermionic emission

nickel. These have a much higher emission efficiency than either tungsten or thoriated tungsten. The working temperature is only 750 °C but because it is easily damaged by ion bombardment, oxide emitting materials are limited to devices working at less than 1 kV.

Review

Check your knowledge and understanding. You should:

Be able to describes the phenomenon of thermionic emission and name materials commonly used for emission in thermionic devices.

5.3 Thermionic diode

The thermionic diode (or **vacuum diode**) is an electronic device with rectifying properties similar to those of the semiconductor diode. It allows current to pass through in one direction, but blocks current from passing in the opposite direction.

The thermionic diode normally comprises two concentric cylindrical **electrodes** called the **anode** and the **cathode**. Both are supported in a glass envelope evacuated of air (Fig. 5.2a). The inner cylinder, the cathode, is made of a suitable emissive material. The outer electrode, the anode, is a normal conducting metal. The British Standard symbol is shown alongside Fig. 5.2a.

The cathode can be heated either indirectly, having a separate heating element inside it, or directly heated by the passage of current through a filament cathode (Fig. 5.2b). Indirect heating is the more popular method. A space charge is formed around the cathode when the operating temperature is reached.

When the anode is set at a more positive potential than the cathode (Fig. 5.3a), by say a battery, electrons from the space charge are attracted through the vacuum to the anode. The diode is then forward biased and the anode current I_A flows in the anode circuit. The anode current shown in Fig. 5.3 is conventional current (which flows in the opposite direction to electron flow). The circuit is completed or 'switched on' only by electrons provided from the space charge.

Conversely, if the anode becomes more negative than the cathode (Fig. 5.3b) no electrons are able to travel to the anode because of electrostatic repulsion. No current flows in the anode circuit and the diode is reverse biased.

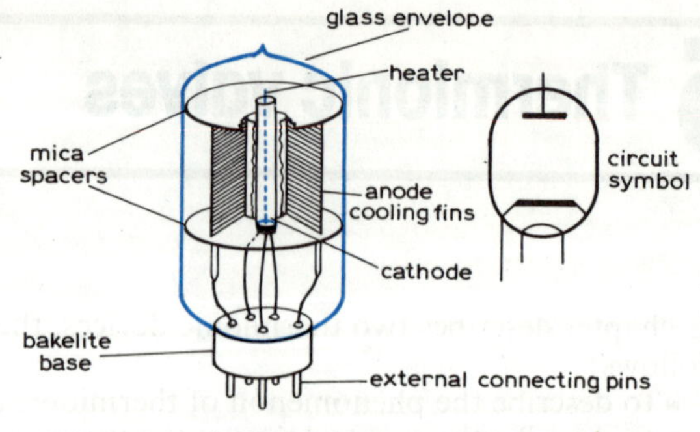

a Indirectly heated cathode

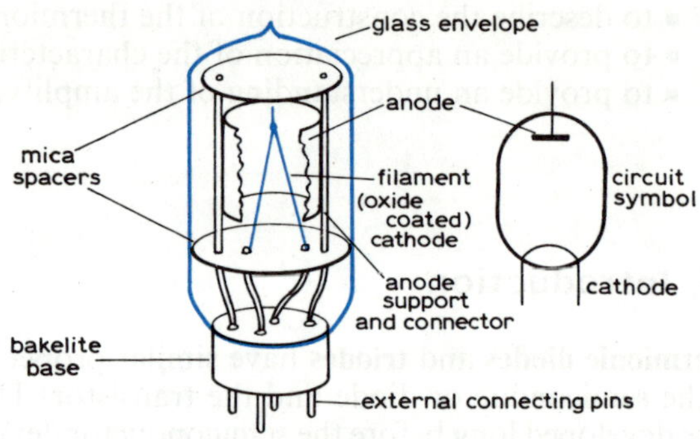

b Directly heated filament cathode

Fig. 5.2 The thermionic diode

5.4 Thermionic diode characteristics

Fig. 5.3a shows a circuit which can be used to obtain the thermionic diode characteristic of anode current I_A against anode voltage V_A shown in Fig. 5.3c. Three regions can be identified from the characteristic; the reverse biased region, the space charge limited region, and the temperature limited region.

The reverse biased region shows that when V_A goes negative with respect to the cathode, no current flows. At zero anode voltage, however, a few electrons attain sufficient energy to make the journey to the anode. This is shown as a very small current when $V_A = 0$.

In **the space charge limited region**, I_A increases as more electrons are drawn from the space charge by the increasing positive V_A.

The temperature limited region exhibits a levelling off of the characteristic. This occurs when all the electrons have been drawn from the space charge which becomes depleted and cannot sustain an increase of current. The steady saturation current I_s is supplied by more electrons being made available by thermionic emission at a constant rate. In this region, the anode current has become independent of V_A and can be increased only by applying more current to the heater to raise the cathode temperature.

5.5 Anode a.c. resistance (or slope resistance)

The thermionic diode characteristic is not linear over its entire range consequently the d.c. resistance V_A/I_A changes with anode voltage. The linear part of the characteristic is used to provide a constant measure of the current and voltage relationship brought about by small variations of anode voltage. The ratio $\Delta V_A/\Delta I_A$ (Fig. 5.3c) is called the **anode a.c. resistance** (or the **anode slope resistance**) r_a

$$r_a = \frac{\Delta V_A}{\Delta I_A} \text{ ohm} \qquad (5.1)$$

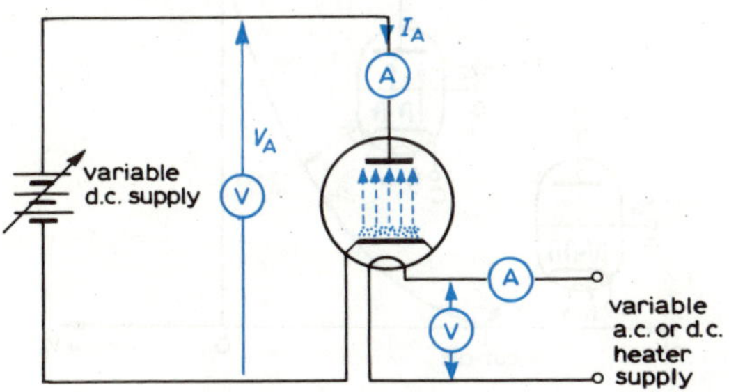

a Circuit for measuring characteristic

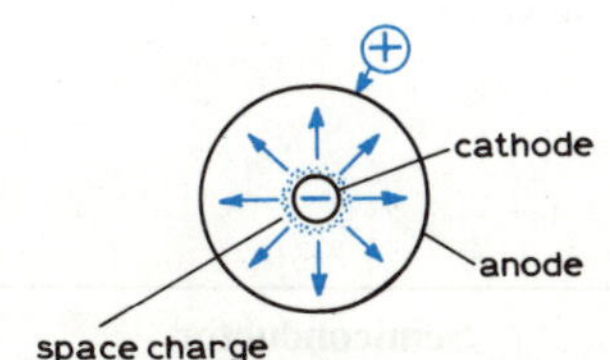

b Forward and reverse biasing

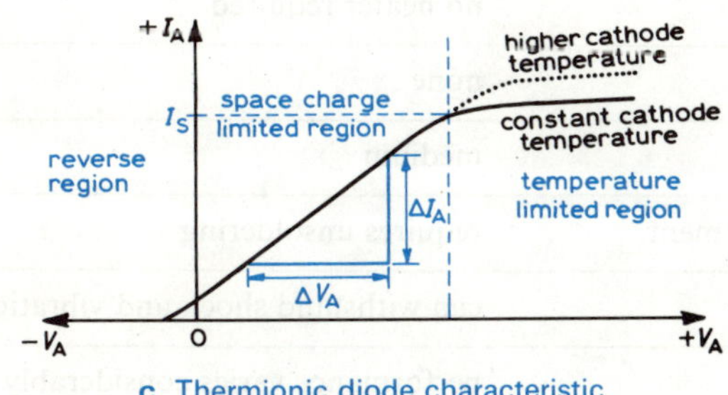

c Thermionic diode characteristic

Fig. 5.3 The action of the thermionic diode

Review

Check your knowledge and understanding. You should:

- **a** Be able to sketch a test circuit diagram for determining the anode characteristic of a thermionic diode.
- **b** Appreciate how current flow is produced in the thermionic diode by correct application of a p.d. between the anode and the cathode.
- **c** Know how to plot and describe a typical anode characteristic.
- **d** Understand the effect of saturation in relation to anode characteristics.
- **e** Appreciate the effect on anode characteristics of a change in heater current.

Exercise

5.1 Identify which of the following is *not* an emissive material:

a tungsten **c** thoriated tungsten
b copper **d** oxide

5.2 Identify the following statements as true or false:

- **a** Thermionic emission is the release of electrons by heat.
- **b** Space charge is the accumulation of positive charge within an emissive material.
- **c** Electrons flow from anode to cathode when the thermionic diode is reverse-biased.
- **d** Thermionic diode characteristics have three distinct regions: the reverse-biased region, the space charge limited region and the temperature limited region.

5.3 Determine which of the following conditions would reverse-bias a vacuum diode:

- **a** Anode voltage = +10 V; cathode voltage = 0 V
- **b** Anode voltage = −10 V; cathode voltage = −20 V
- **c** Anode voltage = 0 V; cathode voltage = −10 V
- **d** Anode voltage = −10 V; cathode voltage = −5 V

5.4 Referring to the linear portion of the I_A/V_A characteristic of a diode valve the anode voltage increases from 75 V to 129 V to increase the anode current from 12 mA to 22 mA. Calculate the a.c. or slope resistance of the diode:

a 6400 Ω **b** 5400 Ω **c** 4600 Ω **d** 3700 Ω

5.5 The slope of the linear part of the I_A/V_A characteristic of a diode valve is 1.92 mA/9.98 V. Calculate the a.c. or slope resistance:

a 5200 Ω **b** 8000 Ω **c** 2500 Ω **d** 4650 Ω

5.6 Comparison of thermionic and semiconductor devices

A comparison between thermionic and semiconductor devices is given in Table 5.1.

5.7 Thermionic triode

The construction of the thermionic triode is very similar to the diode with the addition of a third electrode. The third electrode, the **control grid**, usually takes the form of a loosely wound spiral of wire positioned between the anode and the cathode (Fig. 5.4a) but has no electrical connection with either.

A negative voltage applied to the grid controls the current through the triode as follows. Consider the situation where the grid voltage is initially zero (Fig. 5.4b), and a constant anode voltage is applied. The electrons pass unimpeded through the spaces in the grid and the triode conducts an anode current in exactly the same manner as the diode.

When the grid voltage is made negative with respect to the cathode, some of the electrons are repelled by the electric field around the grid and prevented from reaching the anode. Only the electrons with higher energies pass through.

Increasingly negative V_G repels more and more electrons, reducing the anode current to the **cut-off point** where it becomes zero. There is no practical advantage in making the grid voltage positive as the grid would then attract electrons and conduct like an anode.

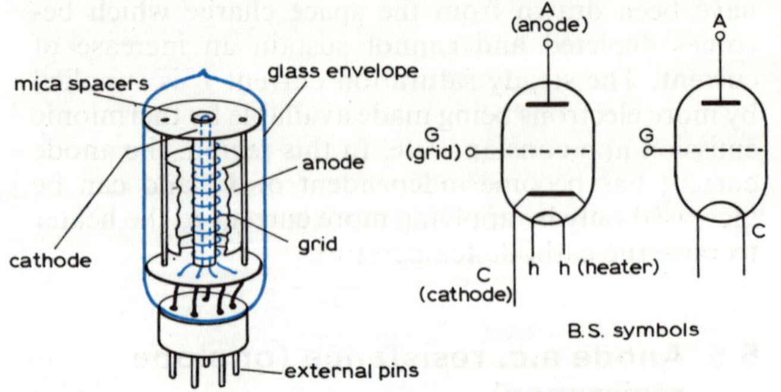

a The triode and BS symbols

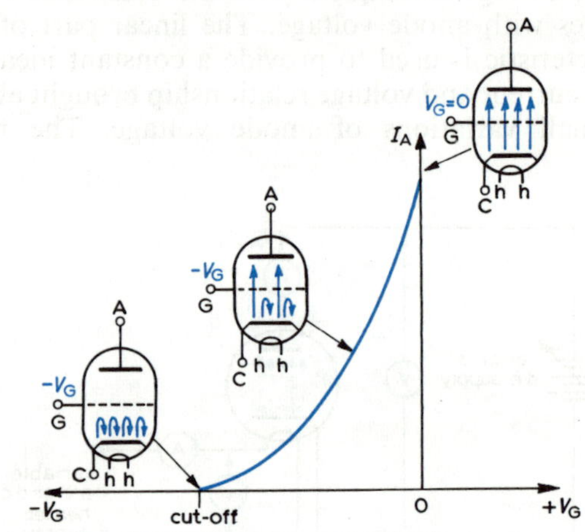

b Transfer characteristic

Fig. 5.4 The triode valve

Table 5.1 Comparison of thermionic and semiconductor devices

	Thermionic	Semiconductor
Size	large	small
Power requirements	requires heater	no heater required
Warm-up time	tens of seconds	none
Maximum power rating	high	medium
Replacement	mounted in holder for easy replacement	requires unsoldering
Robustness	fragile	can withstand shock and vibration
Temperature sensitivity	can withstand high temperatures	performance varies considerably with temperature.

The graph of I_A/V_G shown in Fig. 5.4b is known as the **static mutual characteristic** (or **transfer characteristic**).

The controlling ability of the grid enables the triode to be used for voltage amplification. If a suitable resistor is placed in series with the anode (Fig. 5.5), a small voltage variation at the grid can control a larger voltage change at the anode. The following example emphasises this point.

Example 5.1

A triode has a 10 kΩ series resistor in the anode circuit and is supplied with 240 V d.c. (Fig. 5.5)
When $V_G = 0$ the anode current is 8 mA.
When $V_G = -4$ V the anode current is 2 mA.
By how much does the anode voltage change?

When $V_G = 0$, the 240 V d.c. supply is distributed across the triode and resistor as follows:

Potential difference across $R_L = I_A \times R_L$
$= 8 \times 10^{-3} \times 10 \times 10^3$
$= 80$ V

P.D. across triode $= 240 - 80$
$= 160$ V

When $V_G = -4$ V, the supply is distributed as follows:

P.D. across R_L $= 2 \times 10^{-3} \times 10 \times 10^3$
$= 20$ V

P.D. across triode $= 240 - 20$
$= 220$ V

For a grid voltage swing of -4 V, the anode potential has changed by $220 - 160 = 60$ V, a voltage gain of $60/4 = 15$.

If an alternating signal is applied to the grid causing a variation of ΔV_G, an amplified representation of the signal would appear at the anode (Fig. 5.5). As mentioned earlier, it is undesirable for the anode to become positive with respect to the cathode. So, to prevent the positive half cycle of the signal taking V_G positive, a negative grid bias voltage $-E_B$ is applied, and the signal adds or subtracts from this d.c. value. The value of $-E_B$ must have a magnitude greater than the peak of the signal.

Note that in Example 5.1 that as the grid voltage becomes more negative, the anode voltage becomes more positive and *vice versa*. When amplifying a continually changing signal, the triode amplifier produces an inverted output, (one that is 180° out of phase with the input). As it is mainly the amplitude and frequency that carries the desired information (such as speech or music to a loudspeaker), the phase change does not matter.

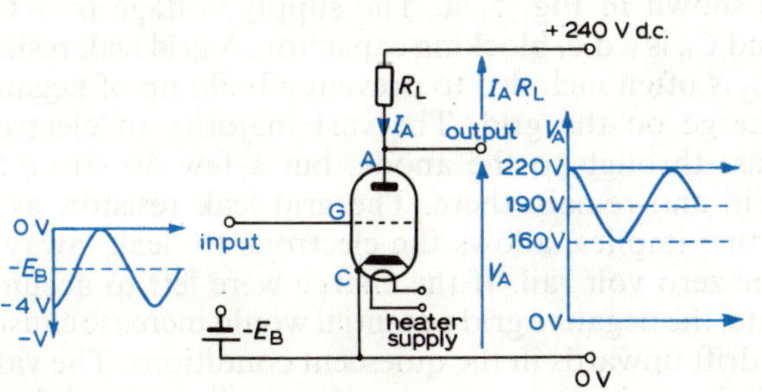

Fig. 5.5 Triode voltage amplification

5.8 Triode characteristics

A circuit for obtaining the triode characteristics is shown in Fig. 5.6a. The graph of I_A/V_G has already

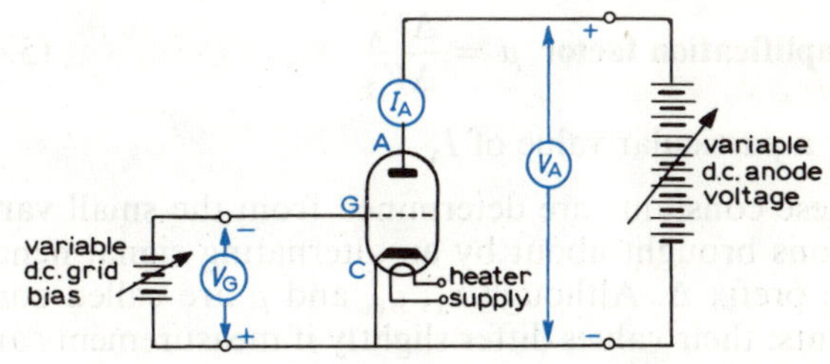

a Circuit for measuring characteristics

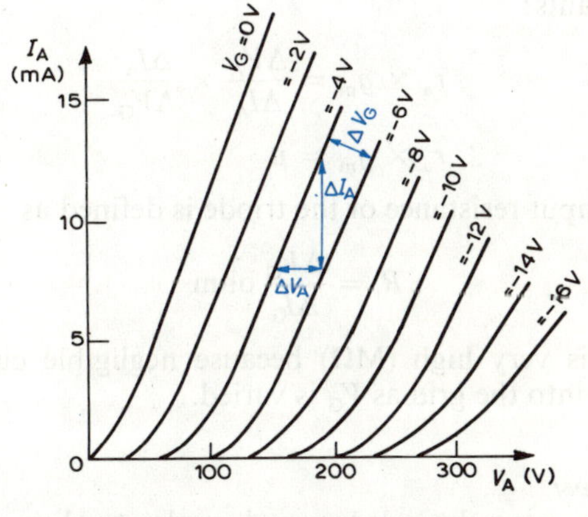

b Triode output characteristics

Fig. 5.6 Measuring output characteristics a b

been described (Fig. 5.4b) and is called the **triode transfer characteristic**.

A family of **output characteristics** provides important information about the device (Fig. 5.6b). When V_A is zero, the grid has no effect and the graph of I_A/V_A is the same as that for the thermionic diode. As V_A is made increasingly negative in fixed steps, the characteristic is moved progressively along the horizontal axis to the right, or less I_A for the same value of V_A.

Three useful constants can be defined from the characteristics which help us to evaluate a particular triode's performance and compare it with others: the anode (or slope resistance), mutual conductance and amplification factor.

Anode (or **slope resistance**)

$$r_a = \frac{\Delta V_A}{\Delta I_A} \text{ ohm} \quad (5.2)$$

r_a is specified for a particular value of V_G.

Mutual conductance

$$g_m = \frac{\Delta I_A}{\Delta V_G} \text{ siemen} \quad (5.3)$$

for a particular value of V_A.

Amplification factor $\mu = \dfrac{\Delta V_A}{\Delta V_G}$ (5.4)

for a particular value of I_A.

These constants are determined from the small variations brought about by an alternating signal hence the prefix Δ. Although r_a, g_m and μ are called constants, their values differ slightly if measurements are taken from different parts of the characteristics.

A relationship can be observed between the three constants:

$$r_a \times g_m = \frac{\Delta V_A}{\Delta I_A} \times \frac{\Delta I_A}{\Delta V_G}$$

$$\therefore r_a \times g_m = \mu \quad (5.5)$$

The input resistance of the triode is defined as

$$R_i = \frac{\Delta V_G}{\Delta I_G} \text{ ohm} \quad (5.6)$$

This is very high (MΩ) because negligible current flows into the grid as V_G is varied.

Review

Check your knowledge and understanding. You should:

a Be able to sketch the circuit symbol of a triode.

b Know the effect of control grid voltage on anode current.

c Be able to sketch a circuit diagram for determining the static characteristics of a triode.

d Understand the static characteristics of a triode and be able to sketch typical families of curves of I_A/V_A (output characteristics) and I_A/V_G (transfer characteristics).

e Appreciate that the input resistance of a triode is high.

f Understand the definitions of anode slope resistance (r_a), mutual conductance (g_m) and amplification factor (μ) and be able to state and derive the relationship between r_a, g_m and μ.

g Know how to obtain the values of r_a, g_m and μ from triode characteristics.

Exercise

5.6 Identify the following statements as true or false:

a The grid of a triode is used to control the flow of electrons.

b Current flows in the grid when a signal is present.

c The triode can be used as a diode if the grid is disconnected.

d The triode amplifier provides a voltage gain with indeterminate current gain.

5.7 Identify which of the following answers completes the statement:

The triode output characteristics show a

a graph of V_A against I_G with constant V_G

b graph of I_A against V_A with constant V_G

c graph of V_G against V_A with constant I_A

d graph of V_A against V_G with constant I_A

5.9 Triode amplifier load line

A basic practical single-stage triode amplifier circuit is shown in Fig. 5.7a. The supply voltage is $+V_{HT}$ and C_0 is a d.c. blocking capacitor. A **grid leak resistor** R_G is often included to prevent a build up of negative charge on the grid. The vast majority of electrons pass through to the anode, but a few do strike the grid and remain there. The grid leak resistor, as its name implies, allows the electrons to 'leak' away to the zero volt rail. If the charge were left to accumulate, the negative grid potential would increase causing a drift upwards in the quiescent conditions. The value of R_G is high to prevent the applied signal being drained away, short-circuiting the triode.

From Fig. 5.7a the following voltage relationship can be deduced:

$$V_{HT} = I_A R_L + V_A \quad (5.7)$$

Rearranging for I_A:

$$I_A = \frac{-V_A}{R_L} + \frac{V_{HT}}{R_L} \quad (5.8)$$

Eqn. 5.8 is a straight line equation of the form $y = mx + c$. There is a similarity here to the d.c. load line for the bipolar transistor amplifier (Section 4.4). Indeed, a load line can be constructed on the triode output characteristics (Fig. 5.7b), which gives all the possible variations of I_A and V_A along a line whose slope is $-1/R_L$ for a given set of circuit conditions (V_{HT} and R_L).

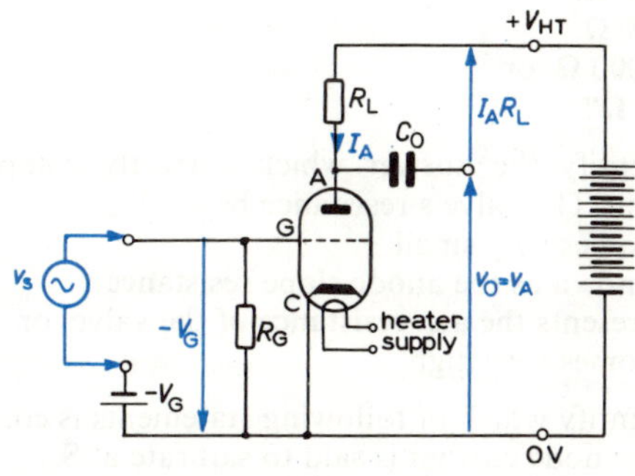

a Simple biasing

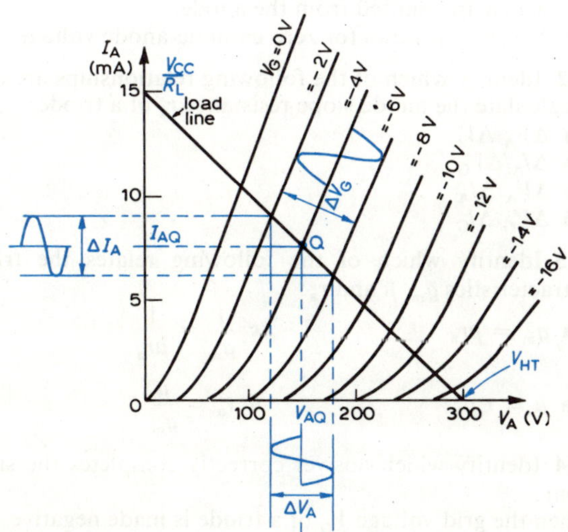

b Load line on characteristics

Fig. 5.7 A triode amplifier

Inserting in Eqn. 5.8 the conditions when $V_A = 0$ and then $I_A = 0$ we get the two end points of the load line:

$$I_A = \frac{V_{HT}}{R_L} \text{ and } V_A = V_{HT}$$

The quiescent values (V_{AQ} and I_{AQ}) are determined by supplying a bias voltage $-V_G$, which fixes the quiescent point Q on the load line. An input signal voltage v_s causes a variation about $-V_{GQ}$. The corresponding changes of I_A and V_A can then be measured from the characteristics.

$$\text{The voltage gain} \quad A_v = \frac{\Delta V_A}{\Delta V_G} \quad (5.9)$$

This is a ratio and therefore dimensionless.
For a sinusoidal waveform, the r.m.s. value of the output voltage is $\Delta V_A/2\sqrt{2}$.

5.10 Automatic biasing of the triode amplifier

The effect of thermal runaway (described in Section 4.6) occurs only in transistors, but **automatic biasing** is still necessary for the triode in order to fix a quiescent point. If the quiescent point is not held stable (in a central position on the characteristics, allowing optimum signal swing) the output might become distorted (as with the transistor amplifier).

Providing a separate negative grid bias battery V_G is cumbersome so a cathode resistor R_K (Fig. 5.8) is employed; quiescent current flowing through the triode passes through R_K raising the cathode potential to $+V_K = I_K R_K$. Therefore, the grid is negative with respect to the cathode because it is clamped to zero

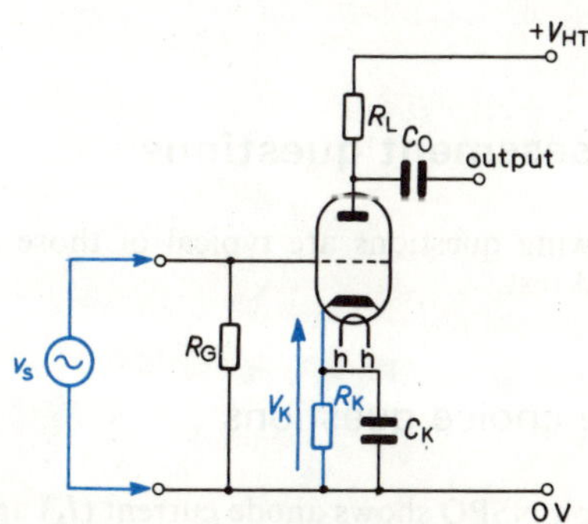

Fig. 5.8 Automatic biasing of triode valve

volts by the grid leak resistor R_G. The anode voltage is still positive with respect to the cathode.

The quiescent point is held constant in the centre of the load line as follows. If I_{AQ} increased for any reason, V_K would be increased, decreasing the potential between grid and cathode. This reduction would lower the anode current, tending to compensate for any variation.

A by-pass capacitor C_K is included to prevent the signal being dissipated in R_K as maximum output voltage is required across R_L.

Review

Check your knowledge and understanding. You should:

- **a** Be able to sketch the circuit diagram of a single-stage triode voltage amplifier having a load resistor, with all direct voltages and currents indicated.
- **b** Appreciate that bias is required to obtain a selected quiescent operating point on the output characteristics.
- **c** Understand the effect of a small sinusoidal voltage input on the quiescent conditions.
- **d** Know that voltage phase inversion occurs between input and output signals.
- **e** Be able to construct the load line for a stated value of load resistance on a given set of output characteristics of a triode amplifier and select a suitable value of bias voltage to obtain a desired quiescent point on the output characteristics.
- **f** Know how to estimate the r.m.s. voltage output from the load line for given quiescent conditions and input signal.
- **g** Know how to determine the voltage gain A_v of the stage.
- **h** Be able to explain, with the aid of sketches, simple methods of biasing a triode amplifier stage.

Self assessment questions

The following questions are typical of those asked in an assessment test.

Multiple choice questions

• The graph NSPQ shows anode current (I_A) against anode voltage (V_A) for a diode valve. Questions 5.8 to 5.11 refer to this graph.

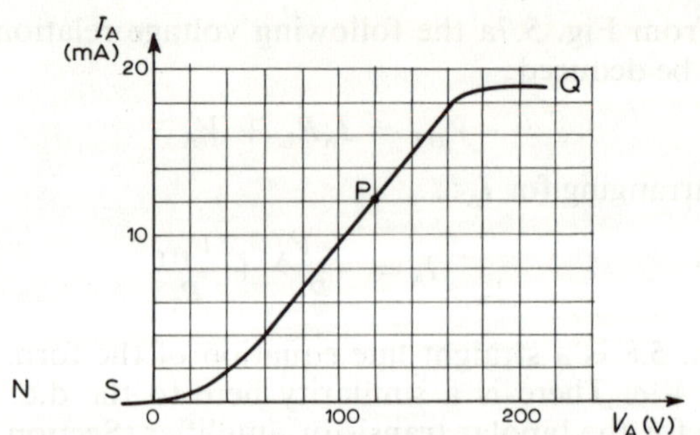

5.8 Calculate the valve's d.c. resistance at P, is it
- **a** 10 Ω,
- **b** 1000 Ω,
- **c** 10,000 Ω, or
- **d** 100,000 Ω?

5.9 Calculate the anode slope resistance at P, is it
- **a** 1000 Ω,
- **b** 7500 Ω,
- **c** 10,000 Ω, or
- **d** 750 Ω?

5.10 Identify the answer which correctly completes the statement: The valve's resistance beyond Q.
- **a** becomes very small
- **b** is known as the anode slope resistance
- **c** represents the a.c. resistance of the valve, or
- **d** becomes very high

5.11 Identify which of following statements is correct:
- **a** The anode current is said to saturate at S.
- **b** At saturation all the electrons escaping from the cathode cross to the anode.
- **c** The current in the region of P is carried by electrons which are emitted from the anode.
- **d** No current flows for zero cathode-anode voltage.

5.12 Identify which of the following relationships are used to calculate the anode slope resistance r_a of a triode:
- **a** $\Delta V_A/\Delta V_G$
- **b** $\Delta I_A/\Delta V_G$
- **c** $\Delta V_A/\Delta I_A$
- **d** $\Delta V_G/\Delta I_G$

5.13 Identify which of the following relates the triode characteristics g_m, μ and r_a:
- **a** $g_m = \mu r_a$
- **b** $\mu = r_a g_m$
- **c** $\frac{1}{g_m} = \frac{1}{\mu r_a}$
- **d** $r_a = \frac{\mu}{g_m}$

5.14 Identify which answer correctly completes the statement:
When the grid voltage V_G of a triode is made negative
- **a** electrons are accelerated
- **b** current flows in the grid circuit
- **c** the anode current is unaffected
- **d** the flow of electrons from cathode to anode is controlled

Short answer questions

5.15 Plot the following vacuum diode characteristics and determine the anode a.c. resistance for an anode voltage of 115 V.

V_A	(V)	25	50	75	100	125
I_A	(mA)	0.15	0.8	1.9	3.7	6.2

5.16 Plot the following output characteristics of a triode valve. By taking measurements from the centre of the characteristics determine

a the anode slope resistance r_a,
b the mutual conductance g_m, and
c the amplification factor μ.

Independently verify the valve of μ from answers r_a and g_m. Output characteristics data is shown in the table below:

5.17 Using the characteristics in question 5.16 determine the values of r_a and g_m at the following points:
a $V_A = 100$ V and $V_G = 1$ V
b $V_A = 250$ V and $V_G = -5$ V
Using the values of r_a and g_m calulate the value μ.

5.18 The triode whose output characteristics are tabulated in question 5.16 is to be used in the circuit of Fig. 5.7a. where $+V_{HT} = 200$ V and $R_L = 40$ kΩ

a Plot the load line on the output characteristics.

b Choose a suitable quiescent point and determine I_{AQ} and V_{AQ}

c If a voltage signal v_s of 2 V peak-to-peak is applied at the input, calculate the voltage gain.

d Describe what would happen to the output signal if v_s increased the grid voltage above zero.

V_A(V)	V_G(V)								
	0	−1	−2	−3	−4	−5	−6	−7	−8
	I_A(mA)								
0	0								
25	1.1	0.3							
50	2.6	1.0	0.2						
75	4.4	2.0	0.7	0					
100	6.3	3.4	1.5	0.3					
125		5.3	2.7	0.9	0.1				
150			4.3	2.0	0.5	0			
175			6.2	3.4	1.3	0.3			
200				5.0	2.5	0.8	0.1		
225				6.8	4.0	1.7	0.5	0	
250					5.6	3.0	1.3	0.3	
275						4.5	2.3	0.9	0.1
300						6.2	3.7	1.9	0.6

6 Cathode ray oscilloscope

This chapter looks at one of the most important instruments for analysing and designing electronic devices – the cathode ray oscilloscope. The aims of this chapter are as follows:

- to describe the construction and principle of the cathode ray tube
- to describe the function of the control circuitry to be connected to a CRT to provide a cathode ray oscilloscope.

6.1 Introduction

The CRO is a device which enables electrical signals to be observed in a graphical form on a screen and for measurements of the signals to be taken. The signals are displayed on a **cathode ray tube (CRT)**. A special form of CRT provides the picture tube in a television.

6.2 Elements of the cathode ray tube (CRT)

Fig. 6.1a shows the essential elements of a CRT. Its electrodes and plates are contained in an **evacuated glass enevelope**, at the end of which is a screen coated with a phosphor. At the other end of the tube is an electron gun which focuses a thin beam of electrons onto the screen. At the beam's point of impact the phosphor glows causing a point of light to be visible when the screen is viewed.

If the position of the electron beam is swept from left to right across the screen (x-axis) at regular intervals (controlled by a **time-base** or **horizontal sweep**), the trace appears as a continuous line. This is because the phosphor has a sufficient **persistence of glow**.

If a time-varying signal voltage can deflect the beam in the vertical direction (y-axis), this combined with regular X deflection provides a display on the screen representing the signal with respect to time.

The regular X deflection is generated by special electronic circuitry called the time base. The signal to be observed as the Y deflection usually needs to be amplified before it can be applied to the Y-plates.

The CRO will be considered in more detail later in

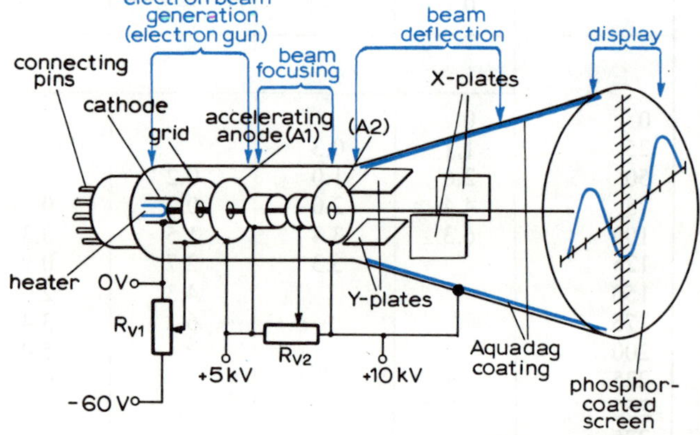

a The basic elements of the cathode ray tube

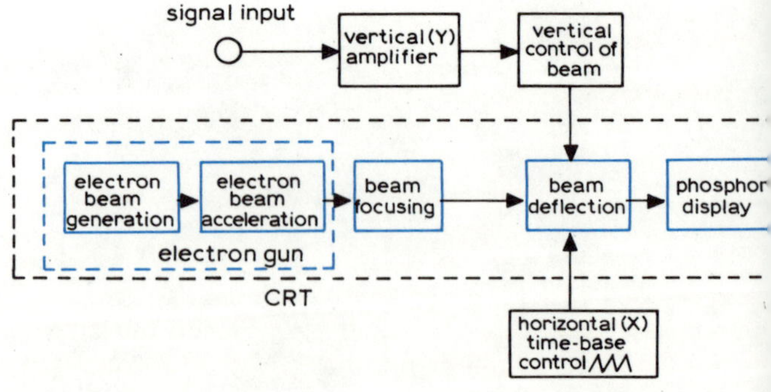

b Block diagram of a cathode ray oscilloscope

Fig. 6.1

this chapter, but first it is necessary to examine the operation of the CRT in more detail. The CRT has four main parts – an electron gun, electron beam focus, deflecting plates and a display screen. These are considered in the following four sections.

6.3 Electron gun

The main components of the electron gun (heater, cathode, grid and anode) follow the same basic principles as in the vacuum triode (Section 5.7). That is, thermionic emission releases electrons which are then accelerated by a high positive potential (with respect to the cathode) applied to anode A_1. The grid carries a negative bias which is variable externally (R_{v1} in Fig. 6.1a). This controls the flow of electrons which determines the intensity of the display (the image is brighter for a greater density of electrons striking the phosphor). After passing through a focusing device, the electrons are accelerated further by anode A_2 which also carries a positive potential. Some of the electrons are collected by the anodes, but many pass through the small holes as a beam.

If negative charge was allowed to accumulate near the screen, the velocity of the electron beam would be reduced and the image would be degraded. This is avoided by quickly removing electrons from the screen area after striking the display by means of a conducting layer of graphite called **aquadag**. This is a coating on the inner surface of the tube (Fig. 6.1a) which is connected electrically to the potential of anode A_2. It not only completes the electrical circuit for the electrons but also has the effect of accelerating the beam after deflection.

6.4 Focusing the electron beam

The electrostatic repulsion between the electrons tends to make the beam diverge causing a patch of light on the display rather than the desired bright spot. To display a fine point on the screen, the beam is focused by an arrangement which may be regarded as an **electron lens**. The result is similar to focusing a beam of light. Fig. 6.2a shows an electrostatic method of focusing the beam. It consists of two tubes A and B containing an electric field. If a higher positive voltage is applied to focusing tube B than to tube A, the resulting field lines are of the pattern shown. The negatively charged electrons travel parallel to the lines. In addition they are accelerated by the electric field emerging as a much narrower, densely packed beam. The lens has the effect of squeezing them through a small aperture. Fine adjustment of the focus is achieved by an externally variable potential difference applied between the focusing tubes.

There is also a method of **electromagnetic focusing** (Fig. 6.2b). The electrons are passed through the centre of a solenoid and are thus deflected by the magnetic field towards the same point on the central axis. This point can be made to coincide with the screen by varying the current through the solenoid until focusing is achieved. This method is not often employed with the CRO but is sometimes used in the television tube.

6.5 Beam deflection

Moving the beam to a desired position on the screen can be done either by electrostatic deflection, or electromagnetic deflection.

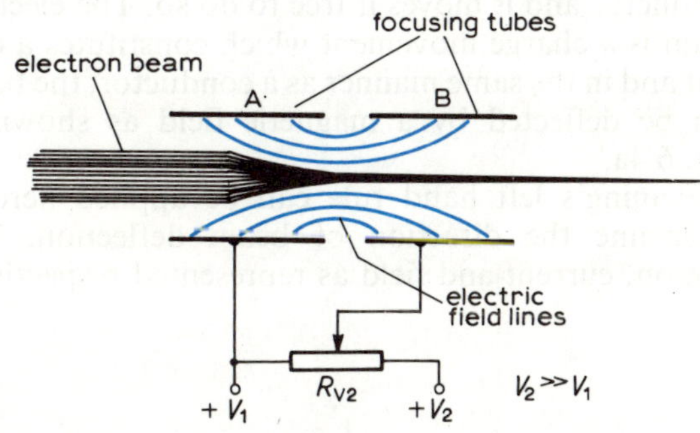

a Electrostatic focusing

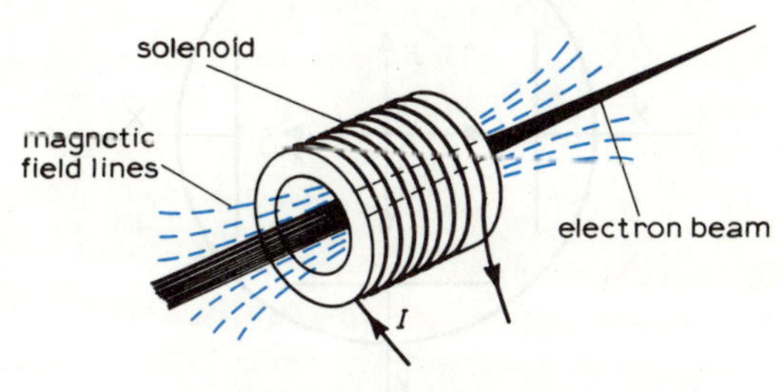

b Electromagnetic focusing

Fig. 6.2

Electrostatic deflection: The plates used inside the tube have sealed external electrical connections to preserve the vacuum. When a substantial potential difference is applied between the plates, the negatively charged beam bends towards the plate which carries the positive potential. If the polarity is reversed, the beam is attracted towards the opposite plate. The angle of deflection depends upon the magnitude of the potential difference between the plates. That is, increase the potential difference and the spot moves further from the centre of the screen (its normal position).

Two sets of plates are required for full control of the beam position (Fig. 6.3). One pair, **the X-plates**, is used to deflect the beam along the x-axis. The other pair, **the Y-plates**, controls movement along the y-axis. (*Note* from Fig. 6.1 that the X- and Y-plates are separated along the axis of the beam to prevent interference between horizontal and vertical control.)

The shape of the plates is carefully designed to give maximum sensitivity in controlling the beam with optimum angle of deflection.

Electromagnetic deflection utilises the force provided by a magnetic field (Fig. 6.4). You may recall from previous study that if a current-carrying conductor is situated in a magnetic field, a force is exerted on the conductor and it moves if free to do so. The electron beam is a charge movement which constitutes a current and in the same manner as a conductor, the beam can be deflected by a magnetic field as shown in Fig. 6.4a.

Fleming's left hand rule can be applied here to determine the direction of beam deflection. The motion, current and field as represented respectively

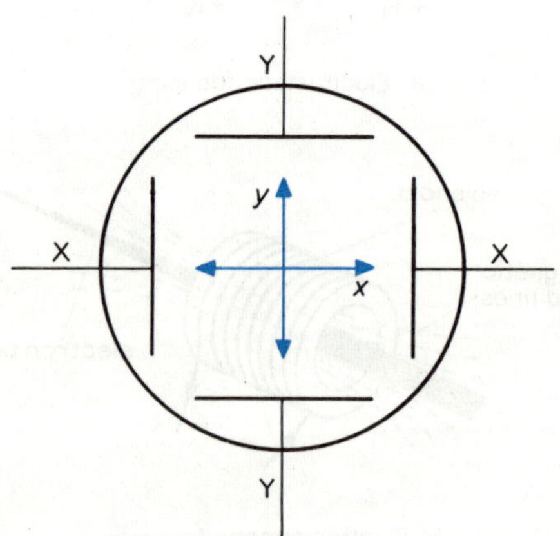

Fig. 6.3 Electrostatic beam deflection

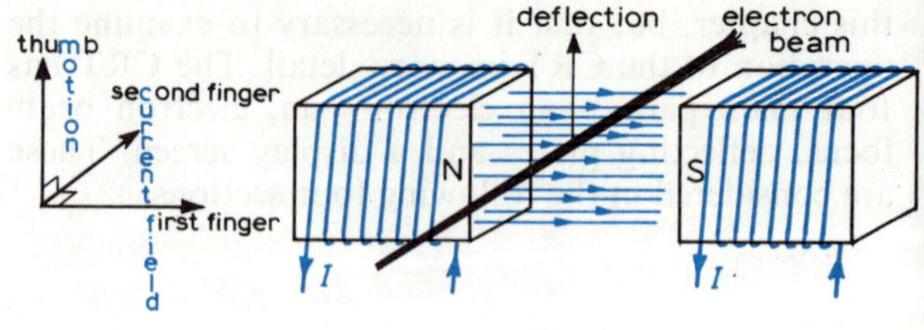

a Flemings left-hand rule

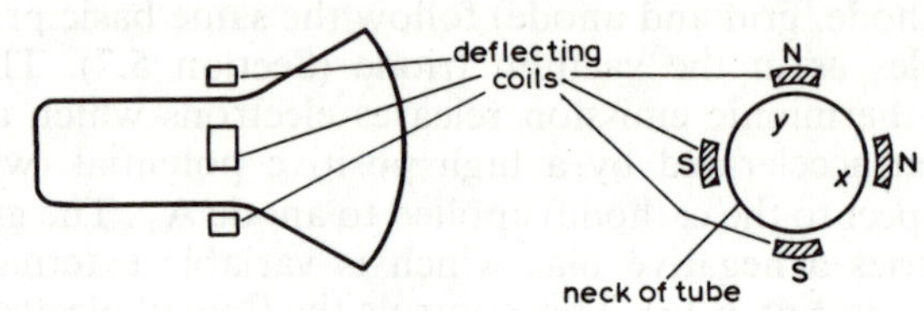

b Position of deflection coils

Fig. 6.4 Electromagnetic deflection

by the thumb, second finger and first finger of the left hand are perpendicular or at right-angles to each other. An important point to remember here is that the direction of the current refers to conventional current which is opposite to that of electron flow. (Although the beam is assumed to be coming out of the page, the second finger must point into the page.)

The degree of deflection depends on the strength of the magnetic field which can be controlled by varying the magnitude of the current flowing in the coils.

For full horizontal and vertical control, two pairs of coils are required (Fig. 6.4b), the X-deflection coils and the Y-deflection coils. Unlike the plates for electrostatic deflection, the coils can be placed radially around the neck of the tube on the outside. This enables a shorter tube length, which is particularly useful for television tubes.

6.6 Phosphorescent display

The display consists of a layer of phosphor which has the property of glowing locally when struck by the electron beam. A thin coating of aluminium is deposited over the phosphor to reflect the light forward through the glass screen.

The time taken for the glow to decay after the beam has moved on is known as the **persistence of glow** (or **afterglow**) of the phosphor. Different phosphors have a range of persistence from below a millisecond to seconds. Each have their own characteristic colour. Many CROs have a green display which is emitted from a medium persistence phosphor. Colour television tubes utilise three types of phosphor which glow respectively red, blue and green (the additive primary colours which are mixed to create any other colour).

Phosphors with different persistence are required for various applications. Some CROs are required to display high frequency signals or occurrences with short time durations. As the trace is repeatedly swept across the screen, it is undesirable for the image of one sweep to interfere with the next. Therefore a phosphor of short persistence is used. For example, if the persistence of a television tube was too long, the image of one frame would be superimposed over the previous frame resulting in a blurring of the picture.

Conversely, some signals to be viewed are of a very low frequency or repeated with long intervals of time between them. For this purpose, the beam is swept slowly across the screen. As it may take a considerable time to repeat the sweep and 're-charge' the glow, a long persistence phosphor is used to preserve the image for as long as possible.

Review

Check your knowledge and understanding. You should:

- **a** Be able to label the diagram of a cathode ray tube.
- **b** Understand the function of the following parts of a CRT; electron gun, focus control, and intensity control.
- **c** Understand how deflection can be produced in a CRT by electric and/or magnetic fields.

Exercise

6.1 Identify which of the following functions are provided by the electron gun in a CRT:

- **a** acceleration of the beam
- **b** deflection of the beam
- **c** focusing of the beam
- **d** shooting electrons into the aquadag coating
- **e** providing a beam of electrons

6.2 Identify the answer which correctly completes the statement:

The electron beam is accelerated by

- **a** the cathode
- **b** the cylindrical anodes
- **c** the X- and Y-plates
- **d** the phosphor coating

6.3 Identify which of the following is *not* used to focus the electron beam:

- **a** optical lens
- **b** electrostatic lens
- **c** electromagnetic lens

6.4 Identify which of the following statements are true and which are false.

- **a** For normal CRO operation, time is represented on the vertical axis, voltage on the horizontal axis.
- **b** The CRT is evacuated, otherwise the cathode will oxidise and burn out.
- **c** If the electron beam was left unfocused, only a blurred patch of light would be displayed.
- **d** The glow of all phosphors decays instantly.

6.7 Scanning control

One sweep of the beam is usually insufficient to display the information adequately. The same waveform must be repeatedly scanned over the screen so that a clear, steady image is displayed. The first step in achieving this is by ensuring that the spot traverses the screen from left to right at regular intervals and with the same velocity. To do this, the sawtooth waveform shown in Fig. 6.5a is applied to the X-plates. As it provides the time axis of the display, this control is called the **time-base**.

To consider the action of the time-base, initially assume that no signal is present at the signal input of the Y-amplifier (Y-input). As the sawtooth ramp voltage increases from value 1 through to 7, the spot on the screen is deflected through points 1 to 7. The amplitude of the sawtooth is made large enough to deflect the spot fully across the screen. The ramp voltage then falls to value 1 which quickly returns the spot to its start position on the left. This is called **flyback**. The action is repeated causing the dot to trace out a straight horizontal line on the display.

The time taken for the sawtooth voltage to change from its most negative value (point 1) to its most positive value (point 7), determines the velocity at which the spot traverses the screen. A steeper slope of the ramp would cause the spot to travel faster. For example, the sawtooth of Fig. 6.5c would result in twice the spot velocity than that of Fig. 6.5a. If

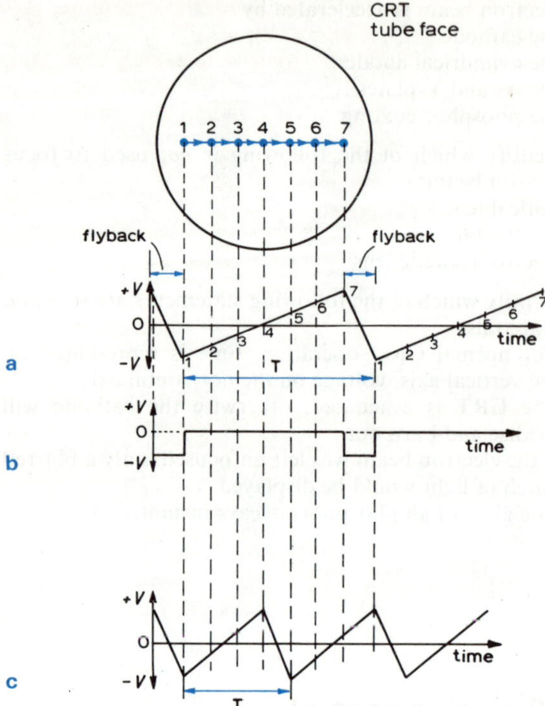

Fig. 6.5 Scanning control
a Time-base sawtooth
b Blanking pulses
c Faster time-base

the amplitude is unchanged, full coverage of the screen is maintained.

The sawtooth is produced by electronic circuitry inside the CRO and can be switched through a range of ramp slopes (time-base control, Fig. 6.1b) giving time-bases from microseconds per centimetre ($\mu s\ cm^{-1}$) of spot motion at the screen to seconds per centimetre ($s\ cm^{-1}$).

Because the flyback period is much shorter than the forward scan period, the beam is moved rapidly back to its start position on the left. As we require to see only the forward trace on the display, the beam is switched off during the flyback period. Another section of the CRO control circuitry applies a negative **blanking pulse** (Fig. 6.5b) to the control grid whilst the time-base ramp potential is returned to point 1.

Now consider the simultaneous application of a time-varying signal on the Y-plates. At the same time as the beam traverses from left to right under the control of the time-base applied to the X-plates, the amplified signal voltage at the Y-plates moves the spot up and down giving a display as shown in Fig. 6.6a. The image is repeated exactly only if the time-base ramp starts its scan (point in Fig. 6.5a) at the same relative point on the input signal cycle.

6.8 Controls of a typical CRO

To the newcomer, the control panel of a CRO can be a bewildering array of knobs and switches. But having read and understood the preceding sections of this chapter you should now be able to deduce the function of some of the basic controls shown in Fig. 6.6a.

The **on/off** and **brilliance** (or **intensity**) are often controlled by the same knob. As you may recall, the intensity is controlled by placing a variable negative voltage to the control grid.

The **focus control** adjusts the voltage applied to the focusing tubes to obtain the sharpest image.

The **X-shift** moves the whole trace left or right by applying a d.c. bias voltage to the time-base sawtooth (Fig. 6.6b).

The **Y-shift** applies a d.c. bias voltage to the Y-plates so that the whole trace can be moved up or down and positioned where required. (Fig. 6.6c).

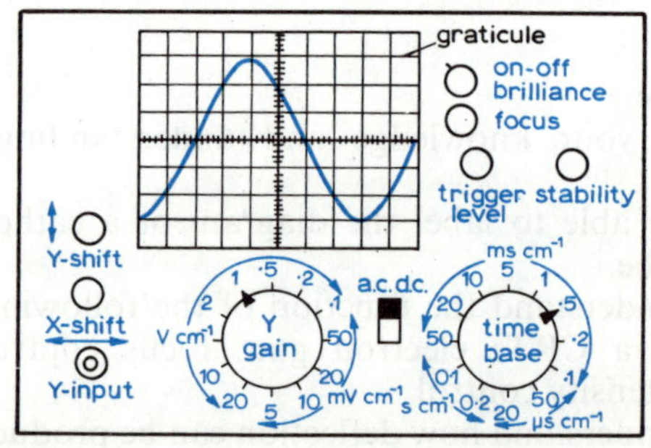

a Controls of a typical CRO

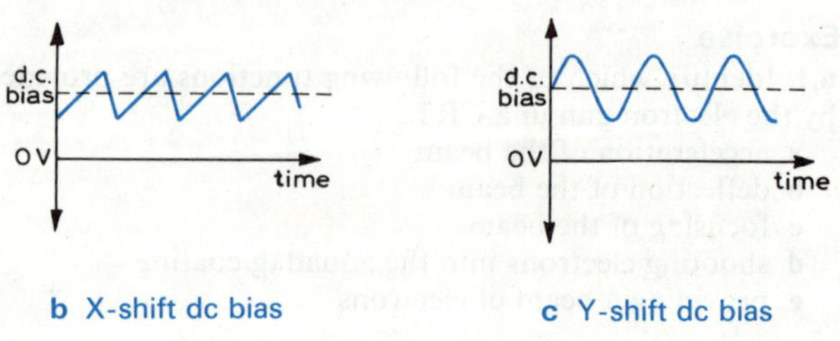

b X-shift dc bias

c Y-shift dc bias

Fig. 6.6

The **Y-gain** determines the gain of the Y-amplifier. Switching this control has the effect of making the displayed amplitude larger or smaller. A selection of scales are available from millivolts per centimetre (mV cm^{-1}) to tens of volts per centimetre (V cm^{-1}) as measured by the **graticule** (a scale grid) in front of the screen.

The **time-base** selects the velocity at which the spot traverses the screen by switching to sawtooth waveforms with the appropriate ramp slope. The range normally extends from μs cm^{-1} to s cm^{-1}.

The **trigger level** sets the predetermined input voltage at which a trigger pulse can activate the time-base. Input sine waves can be used to trigger the time-base on either positive or negative going parts of the wave.

The **stability** control is used to synchronise the time-base frequency with that of the input signal.

The **a.c.–d.c.** switch is used to select the appropriate mode of measurement. Set to a.c., a capacitor is placed in series with the Y-input to cut out any d.c. bias in the signal as this would cause a Y-shift. Switched to d.c., the capacitor is short-circuited so that the d.c. component can be measured directly from the screen.

Review

Check your knowledge and understanding. You should:

a Understand how a regular scan can be obtained in the X-direction of a CRT (the time-base).
b Appreciate the use of blanking pulses in a CRO time-base.
c Be able to describe the use of controls of a CRO.

Exercise

6.5 Identify which part of the following CRO control circuitry has become faulty if the flyback trace can be seen on the display;
a the time-base sawtooth generator
b the trigger circuit
c the synchronisation circuit
d the blanking pulse circuit

6.6 If the positive slope of a time-base sawtooth takes 2 mS to reach a maximum, determine how many cycles of a 3 kHz sinusoidal waveform would be displayed.

6.7 Identify which CRO control is operated to cause each of the following effects:
a The displayed waveform increases in height, but there is no difference in horizontal display.
b The image is moved up and down the screen, but there is no change in image size.
c The trace is originally blurred, becomes clear, then again blurred.
d The displayed waveform contracts along the x-axis, eventually the dot is seen to traverse slowly from left to right.

6.8 Determine the amplitude and frequency of the sine wave displayed in Fig. 6.6 (observe the Y-gain and time-base settings).

Self assessment questions

The following are typical of questions you will be asked in assessment tests:

Multiple choice questions

6.9 Identify the individual components of the CRT by matching the letters **a** to **g** with the corresponding arrow number on Fig. 6.7.
e.g. if you think arrow 6 points to the Y-plates your answer is **g**–6.

a electron gun
b X-plates
c aquadag coating
d phosphor
e focusing tubes
f accelerating anodes
g Y-plates

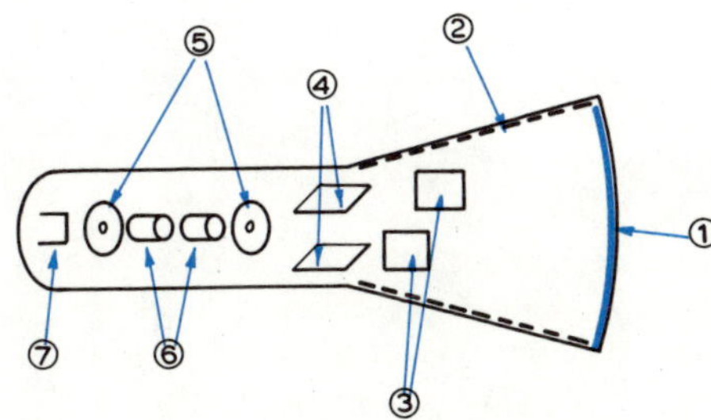

Fig. 6.7

6.10 Identify which of the following components of the CRT is used to vary the intensity of the beam;
a the X-and Y-plates
b the type of phosphor
c the control grid
d a solenoid fixed to the neck of the tube.

6.11 Identify which answer correctly completes the statement.
Blanking pulses are applied to;
- **a** the cathode
- **b** the anodes
- **c** the X- and Y-plates
- **d** the control grid

6.12 Identify which two of the following methods are used to deflect the beam;
- **a** rotating the electron gun
- **b** using an electromagnetic field
- **c** optical reflecting surfaces
- **d** using an electro-static field

6.13 Identify the function of the X-plates;
- **a** to deflect the beam in a vertical direction
- **b** to produce x-rays for the electron beam
- **c** to gather electrons after they have struck the screen
- **d** to deflect the beam in a horizontal direction

6.14 Identify which of the following statements are true and which are false;
- **a** vertical deflection is provided by the Y-plates
- **b** the purpose of blanking pulses is to cut-off the beam during flyback
- **c** the CRO displays a.c. only
- **d** the time-base waveform is applied to the X-plates

Short answer questions

6.15 Describe the components of the electron gun and explain the function of each.

6.16 State the two methods of deflecting an electron beam. Briefly describe each method.

6.17 You require to display a 10 kHz sine wave with an amplitude of 100 mV. Choose a suitable time-base and Y-gain setting from the ranges shown in Fig. 6.6.

7 Oscillators

This chapter describes the basic principles of electronic oscillators. Oscillators are electronic circuits which provide repetitive signals. They are designed to produce waveforms of various shapes, frequencies and amplitudes. The aims of this chapter are as follows:

- to describe the main categories of waveforms and the parameters used to specify them
- to explain the basic principles of oscillators in electrical circuits
- to describe simple practical electronic oscillators

7.1 Oscillator waveforms

There are many types of oscillators designed to give particular output waveforms. These waveforms and their applications are described before discussing the oscillator circuits.

Although many complex waveforms can be generated, we consider three basic categories (Fig. 7.1); sinusoidal waveforms, rectangular waveforms, sawtooth waveforms.

Sinusoidal waveforms: The voltage variation of a sine wave (Fig. 7.1a) follows a distinctive smooth change from peak-to-peak. The time taken for one complete cycle is called the **periodic time** T. The frequency in Hertz is the reciprocal of T.

The following provides some examples the application and occurrence of sine waves in electronics:

a As a signal of constant frequency and amplitude to test the frequency response and gain of amplifiers and other circuits.

b As a carrier wave used in the transmission of radio signals.

c As a reference frequency in the reception of a radio signal.

d As a reference signal for measuring the frequency of other signals.

e Fault tracing by injecting a signal of known frequency and amplitude into the input of a circuit and detecting its presence or absence elsewhere in the circuit.

f As pulses of high frequency for radar.

g Waveform synthesis, where the notes of musical instruments are synthesised by the complex mixture of sine waves of different frequencies.

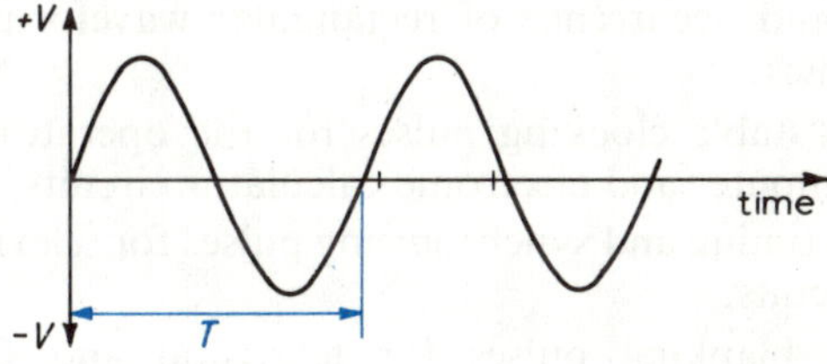

a sinusoidal

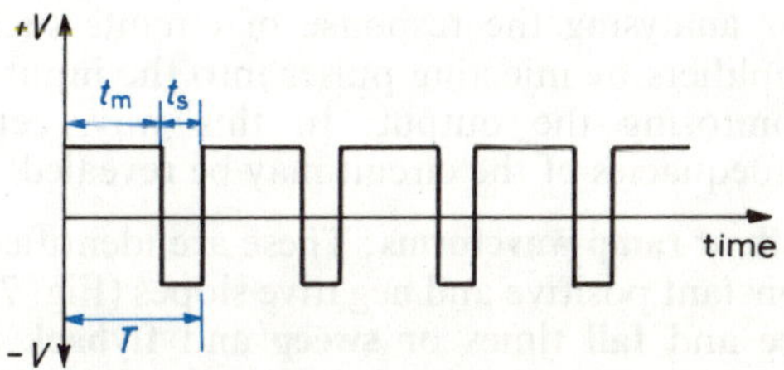

b rectangular

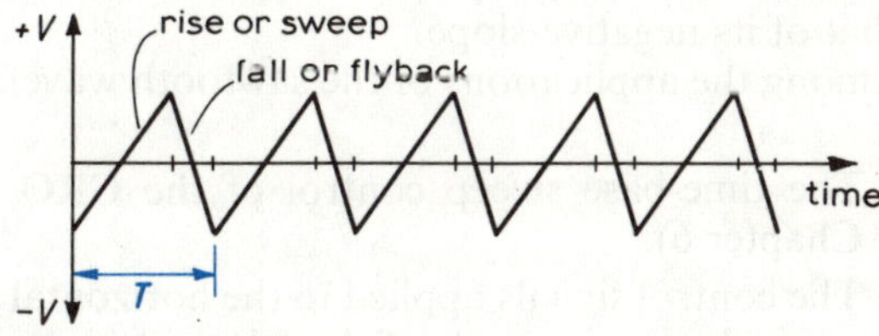

c sawtooth

Fig. 7.1 Basic waveforms

Rectangular waveforms: These are identified by their abrupt change from one voltage level to another (Fig. 7.1b). The change can be from zero to positive and back to zero, or from zero to negative and back to zero, or a change between positive and negative.

Rectangular waveforms have repetitive cycles and so we can define the periodic time *T* and frequency *f*. Further properties of the rectangular waveform can be defined. Unlike the sine wave, the rectangular waveform can have one half cycle of a different time duration than the next, t_m and t_s in Fig. 7.1b. The presence of a pulse is called a **mark**, the absence of a pulse is called a **space**. Hence the ratio of t_m to t_s is termed as the **mark-space ratio**. When the duration of the space is much greater than that of the mark ($t_s \gg t_m$), the number of pulses per second is called the **pulse repetition frequency (p.r.f.)**.

The **square wave** is a special example of the rectangular waveform where the mark duration is equal to the space duration ($t_m \equiv t_s$).

The following provides some examples of the application and occurrence of rectangular waveforms in electronics:

a As stable clocking pulses for the operation of computer and electronic calculator circuits.

b As timing and synchronising pulses for television circuits.

c As blanking pulses for television and CRO circuits.

d For the encoding and decoding of digitally modulated messages.

e For analysing the response of circuits such as amplifiers by injecting pulses into the input and monitoring the output. In this way, certain inadequacies of the circuit may be revealed.

Sawtooth or ramp waveforms: These are identified by their constant positive and negative slopes (Fig. 7.1c). The **rise** and **fall** times or **sweep** and **flyback** may differ significantly. Again the periodic time *T* and frequency *f* are defined.

The **triangular waveform** is a special case of the sawtooth which has the positive slope duration equal to that of its negative slope.

Among the applications of the sawtooth waveform are:

a The time-base sweep control of the CRO (see Chapter 6).

b The control signals applied to the horizontal and vertical scanning coils of the television tube and radar monitors.

c Timing circuits where a fixed duration of time is required for the voltage to reach a predetermined level. Some digital voltmeters use sawtooth waveforms by counting a number of pulses in the time it takes for the ramp to reach the same level as the measured voltage.

Review

Check your knowledge and understanding. You should:

a Be able to sketch and describe the main categories of oscillator waveforms and define the parameters associated with them.

b Know examples of the application and occurrence of sinusoidal, rectangular and sawtooth waveforms in electronics.

Exercise

7.1 Draw the waveform described for each of the following applications. Mark the periodic time and amplitude on each drawing.

a To control the line scanning of a television picture a sawtooth waveform is used where the sweep and fly back times have a ratio of 6 to 1.

b The flyback blanking pulses of a CRO are rectangular with a mark-space ratio of 10 to 1.

c The clock pulses for a particular computer are in the form of a square wave with a periodic time of 10^{-6}s.

7.2 The waveforms of Fig. 7.2 are the result of mixing two basic waveforms. Identify the two waveforms used for each.

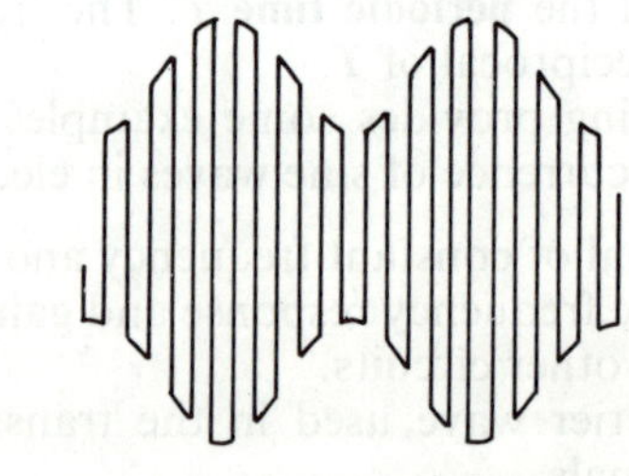

a

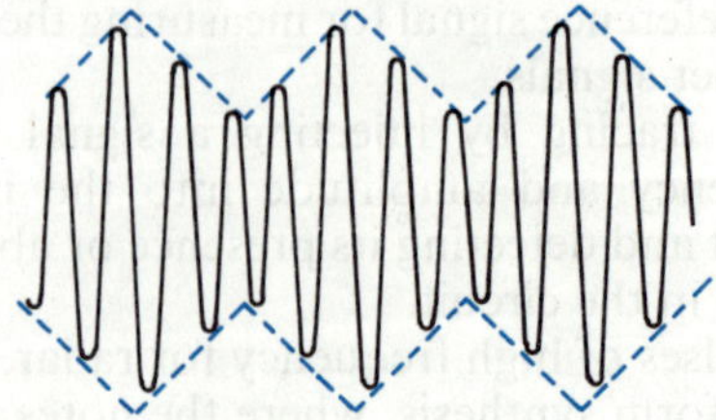

b

Fig. 7.2

7.2 Principles of oscillation

There are two main categories of oscillator; **sinusoidal oscillators** giving pure sinusoidal outputs, and **relaxation oscillators** producing rectangular waveform or pulse outputs.

This chapter is primarily concerned with the sinusoidal oscillators; relaxation oscillators are described in detail in Chapter 14.

Under certain conditions amplifiers become oscillators. Consider Fig. 7.3a where a sinusoidal signal voltage is amplified and a proportion of the output is fed back to the input. When the input signal and the feedback signal are in phase, they are added and the combined signal is amplified. This is known as **positive feedback** (see Chapter 12). The oscillations increase until the amplitude is limited by the circuit elements or the available energy from the power supply. The system continues to oscillate when the input signal is removed provided the feedback signal has sufficient amplitude.

A starting input signal is not needed with a practical oscillator. The initial switching on generates brief or transient oscillations over a wide enough frequency range to trigger the oscillator. This is known as **self starting**.

Sometimes unwanted oscillations are generated in an amplifier system when part of the output signal accidentally finds its way back to the input. You may have heard the high-pitched squeal from the amplifiers of some rock groups. This can occur when the microphone is placed such that it picks up the output from the loudspeakers.

Fig. 7.3b is a block diagram specifying the main requirements of an oscillator which are:

amplification
amplitude limitation
phase inversion of the feedback signal, if necessary,
and **frequency determination**.

The amplitude of oscillations must be limited to give an output of constant amplitude. In simple oscillators this is done by driving a transistor into saturation (See Section 4.6). Additional amplitude limiting circuitry can be manually adjusted to provide an output of variable amplitudes. If the amplitude is too severely limited the feedback may be insufficient to maintain oscillations.

Phase inversion of the feedback signal (Fig. 7.3b) is required if the amplifier itself causes a 180° phase difference between the input and output signals (as with the common-emitter amplifier). This ensures that the input and feedback signals are in phase.

The frequency determining network selects and only passes the frequency of oscillation f_0 to the input, other frequencies are rejected. The frequency of oscillation can also be manually variable.

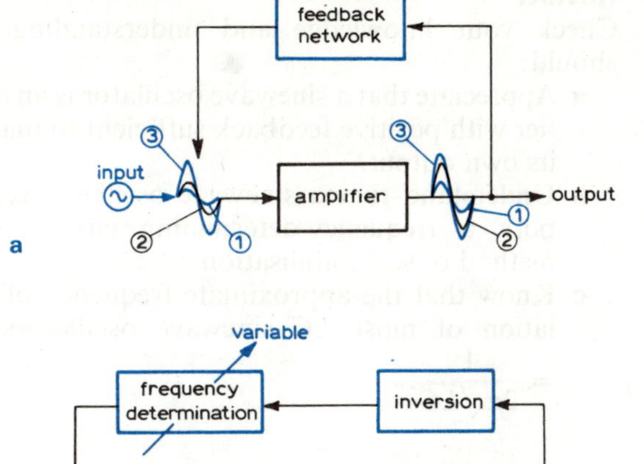

Fig. 7.3 The basic principles of oscillators

To summarise, the essential requirements for stable sinusoidal oscillations are:

a *A resultant zero phase shift around the closed circuit* (i.e. any phase shift through the amplifier is cancelled by the feedback network).

b That the *overall gain around the closed circuit is unity* (i.e. no increase or decrease of the overall signal to ensure undamped oscillations).

The frequency determining circuit is often a parallel combination of an inductor L and capacitor C. The resonant frequency f_r of this circuit is that which makes the reactances equal and therefore cancel each other's effects

i.e. inductive reactance = capacitive reactance

$$X_L = X_C \tag{7.1}$$

$$2\pi f_r L = \frac{1}{2\pi f_r C} \tag{7.2}$$

Rearranging

$$f_r = \frac{1}{2\pi\sqrt{LC}} \tag{7.3}$$

The frequency of oscillation f_o is not exactly equal to f_r because of the additional capacitance of other components in the oscillator circuit and inductive coupling between the leads. Therefore $f_o \approx f_r$

Review

Check your knowledge and understanding. You should:

a Appreciate that a sinewave oscillator is an amplifier with positive feedback sufficient to maintain its own output.

b Understand that a sinewave oscillator requires both a frequency-determining circuit and a method of self-stabilisation.

c Know that the approximate frequency of oscillation of most *LC* sinewave oscillators is $f_o = \dfrac{1}{2\pi\sqrt{LC}}$.

Exercise

7.3 Describe what would happen to the output waveform of an oscillator if the signal energy lost in feedback was not replaced by an amplifier.

7.4 Calculate the frequency of oscillation of an oscillator if the tuned circuit comprises $L = 10$ mH and $C = 100$ pF in parallel.

7.5 Frequency of 320 kHz is required of an *LC* tuned oscillator. Determine which of the following capacitors would you use if $L = 1$ mH.

a 10 μF
b 25 μF
c 200 to 300 pF variable
d 20 to 30 pF variable

7.3 Simple oscillator circuits

A basic inductive-capacitive (*LC*) oscillator circuit is shown in Fig. 7.4a. The tuned circuit is C_1 in parallel with L_1 which is the primary of a transformer. The phase inversion is provided by the transformer windings which are connected so that the secondary voltage is 180° out of phase with the primary voltage.

Fig. 7.4b shows the **tuned-collector oscillator** which is essentially a common-emitter transistor amplifier with a modified collector circuit. Amplitude stabilisation to some extent is provided by resistor R_E which operates in the same manner as the transistor stabilising network described in Section 4.7. Capacitor C_B provides further amplitude stabilisation (it is discussed further in Section 7.4).

The **tuned-anode oscillator** of Fig. 7.4c is a modification of the triode amplifier. The function of its components are similar to the tuned-collector oscillator.

The **R-C network** of Fig. 7.5 provides a feedback

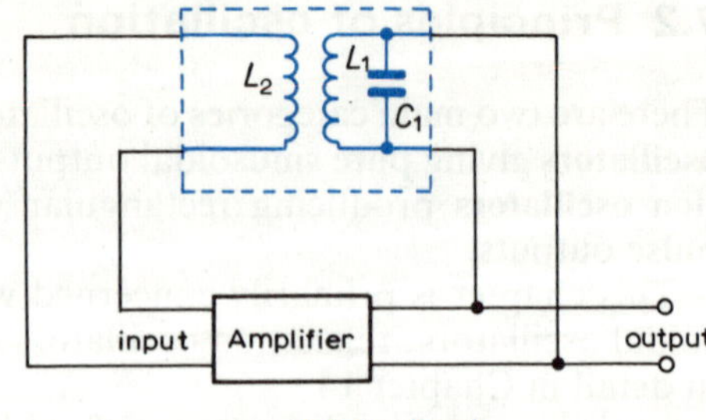

a *LC* feedback

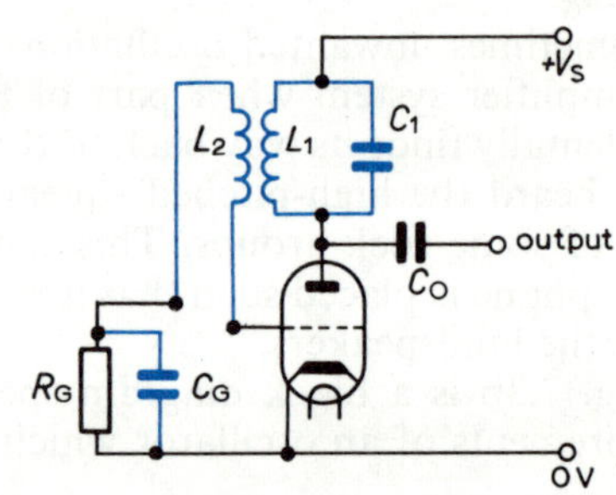

b tuned-collector oscillator

c tuned-anode oscillator

Fig. 7.4 Simple *LC* oscillator circuits

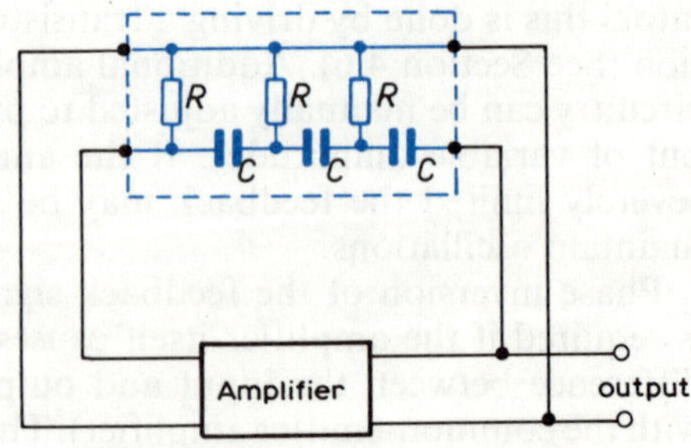

Fig. 7.5 *RC* feedback

path with a 180° phase shift. For sustained oscillation, the amplifier gain must be sufficient to replace the attenuation (reduction of signal) introduced by the *R-C* network.

7.4 Biasing of tuned-circuit oscillators

To ensure the transistor is biased on when the d.c. supply is connected, the potential divider arrangement of R_1/R_2 is provided (Fig. 7.4b). Without capacitor C_B, the transistor would be driven into saturation before the overall gain was limited to unity. (This is referred to as **class A** operation see Section 10.8).

To prevent saturation a refinement is made by the connection of capacitor C_B across R_2 (Fig. 7.4b). This is known as **leaky base biasing**, the action of which is as follows.

On successive negative half cycles of the oscillations, C_B is charged to a negative d.c. value (during the positive half cycles, the accumulated charge cannot leak away completely through R_2). The negative potential on C_B lowers the bias voltage at the base of the transistor so that it is just sufficient to accommodate the peak-to-peak oscillations. Any tendency for the amplitude of the oscillations to increase causes the negative potential of C_B to increase, lowering the base bias voltage even further and hence has a compensating effect on the gain. Any tendency for the amplitude to decrease reduces the negative charge on C_B and thus increasing the gain.

So initially, the biasing is fixed by the potential at the junction of R_1 and R_2. When oscillations occur, the biasing is automatically adjusted by the signal charging C_B to a negative potential. (This is sometimes referred to as **class C** operation, full definitions of classes of biasing are given in Section 10.8).

A similar technique known as **leaky grid biasing** is employed with the tuned-anode oscillator. Hence the inclusion of C_G and R_G in Fig. 7.4c.

Review

Check your knowledge and understanding. You should:

- **a** Be able to sketch the circuit diagrams of a tuned-collector and a thermionic triode oscillator.
- **b** Understand the operation of an *R-C* oscillator.
- **c** Know the methods of applying a bias in a tuned circuit transistor oscillator (class A and C).

7.5 Frequency stabilisation

A main requirement of an oscillator is to generate signals of a stable frequency. In practice variations can occur owing to the following:

- **a** Temperature variations change component values or active device characteristics.
- **b** The active devices may have inherent irregularities or their parameters might vary with age.
- **c** The supply voltage may fluctuate.
- **d** Varying output load conditions can vary the frequency.

One method of improving the stability of an oscillator is to control the frequency using a **piezo-electric crystal**. This is a small crystal (usually quartz) which has the property of mechanically flexing when a voltage pulse is applied to it and, conversely, generating a small voltage when mechanically stressed.

If such a crystal is placed in an oscillator circuit (Fig. 7.6), the mechanical and electrical interaction causes it to vibrate at a natural resonant frequency, the circuit oscillates at the natural frequency with a high degree of constancy.

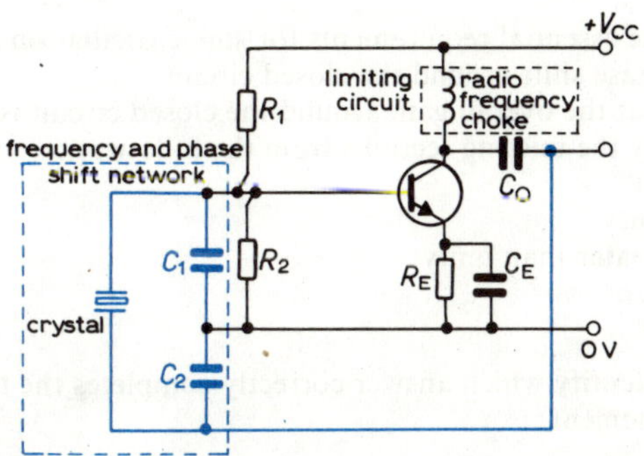

Fig. 7.6 Crystal-controlled oscillator

Review

Check your knowledge and understanding. You should:

- **a** Know the factors which affect the stability of oscillators.
- **b** Understand and be able to describe a method of improving the stability of oscillators using a piezo-electric crystal.

Self assessment questions

The following are typical of assessment test questions.

Multiple choice questions

7.6 Determine which waveforms (sinusoidal, rectangular or sawtooth) are required by the following applications;

a blanking signals for television and CRO circuits
b timing circuit where a fixed duration of time is required for the voltage to reach a predetermined level
c stable clocking pulses for digital circuits
d carrier wave for radio and television signals
e waveform synthesis.

7.7 Identify which answer correctly completes the following statement.
To cause an amplifier to oscillate, the output is fed back with the input

a 270° out of phase
b in phase
c 180° out of phase
d 90° out of phase

7.8 Sketch the elements of a tuned-collector oscillater including the following;

a frequency determining circuit
b stabilising circuit
c amplifier
d inverter

7.9 The essential requirements for stable oscillation are;

i phase shift around the closed circuit
ii that the overall gain around the closed circuit is?

Identify the missing sections from the following;

a 180°
b unity
c greater than unity
d zero
e 90°

7.10 Identify which answer correctly completes the following statement:
The frequency of oscillation f_o of an LC tuned oscillator is determined by

a $f_o = 2\pi\sqrt{LC}$
b $f_o = \dfrac{1}{2\pi\sqrt{LC}}$
c $f_o = \dfrac{1}{2\pi LC}$
d $f_o = \dfrac{2\pi}{LC}$

Short answer questions

7.11 Using the same scales for each, sketch on graph paper the following waveforms;

a a sinusoidal waveform with a peak of 2 V and a periodic time of 10 ms
b a rectangular waveform with a peak of 3 V, a periodic time of 20 ms
c a sawtooth waveform of peak 2.5 V and a sweep/flyback ratio of 3 to 1.

7.12 State two common applications for each of the waveforms.
Sketch the circuit diagram of:

a a tuned-collector oscillator
b a tuned-anode oscillator

Describe their operation.

7.13 In an oscillator explain if you would use an inverting feedback network for;

a a common-emitter amplifier
b a common-base amplifier.

State your reasons why.

7.14 With the aid of an appropriate diagram, describe how leaky base biasing is applied to a tuned-collector oscillator.

7.15 Sketch and label an RC oscillator.

7.16 Describe the causes of frequency instability in oscillators.

8 Digital electronics

This chapter introduces the important subject of digital electronics which is centred around the principle of being able to communicate information by combinations of two-state signals. The aims of this chapter are as follows:

- to describe the difference between analogue and digital signals
- to provide an appreciation of the applications of digital signals
- to provide an understanding of logic levels and their advantages
- to explain two state devices and their applications (AND, OR and NOT gates)
- to describe the operation of simple logic circuits

8.1 Analogue and digital signals

Information can be processed and conveyed by electrical signals. These can be placed in two categories; analogue and digital signals.

Analogue signals: These are signals which vary continually with time. The waveform of an analogue signal, which could be the electrical representation of speech shown on a CRO, is as in Fig. 8.1a. The level and frequency of such a signal always is in proportion to the original signal which it represents (analogue means 'a similar thing'). Electrical analogue signals are used to convey information in microphones, radios, televisions, tape-recorders and so on.

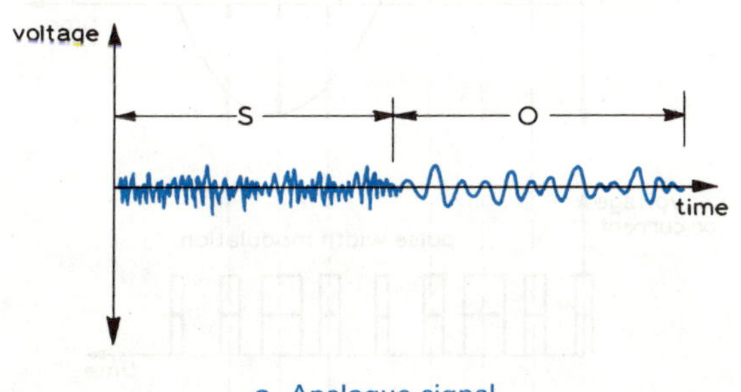

a Analogue signal

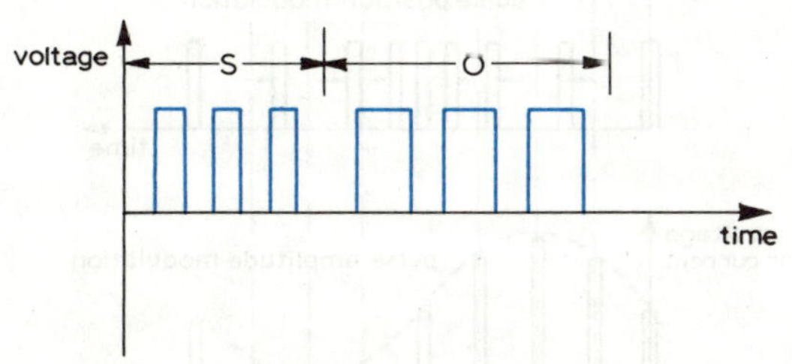

b Binary digital signal

Fig. 8.1

Digital signals: These are signals in the form of **pulses**. In many applications the pulses have two states, ON or OFF. Such signals are called **binary digital signals**. They convey information by being processed through electronic circuits in the form of a pattern which relates to a known code. A widely known example of a binary digital signal in the form of a coded message is expressed in the Morse Code. This code requires ON or OFF signals, their combinations expressing the letters of the alphabet and numbers. Fig. 8.1b shows a graphical representation of the Morse Code binary digital signals for the letters S and O.

There are examples of digital and analogue devices outside the field of electronics. The abacus and slide rule are both devices used to perform calculations. The former is a digital device, the latter analogue.

8.2 Simple applications of two-state signals

Two-state signals which convey messages are part of our every day life. For example, if amber is disregarded, traffic lights have two states; *red* and *green* or *stop* and *go*; many railway signals have two states, *up* and *down*; the brake lights of a motor-cycle are either *on* or *off*; flashing indicators have two states. All these signals are effective methods of conveying information.

At first it might appear that two-state signals have limited use compared with analogue signals. This is not so as highly complex information can be represented by sophisticated coding of two-state signals. In addition to the Morse Code, many digital codes have been devised to serve applications ranging from computers to telecommunications. Many analogue signals, such as the human voice and music, are broken down into pulses, passed through a transmission system and reconstructed at the distant end without the speaker or listener being aware. This method is often used in telecommunication and is called **pulse code modulation (p.c.m.)**.

Two advantages in converting analogue signals are that digital signals deteriorate less when processed and transmitted over distances *and* that low frequency signals, which are difficult to convey and process through systems, can be carried through them on pulses.

Although the binary digital signal has only two states, there are various ways in which the pulses are modified to carry information:

as **coded strings of pulses (p.c.m.)**
by **pulse width modulation (p.w.m.** Fig. 8.2b)
by **pulse position modulation (p.p.m.** Fig. 8.2c)

Another type of pulsed signal utilises the variation of the pulse amplitudes in accordance with the speech or music waveform (Fig. 8.2d). This is called **pulse amplitude modulation (p.a.m.)** in which the pulses have infinitely more than two states. In fact, p.a.m. is really a combination of an analogue and digital signal. Using p.a.m. many conversations or messages can be interlaced by encoders and simultaneously transmitted along the same cable. At the receiving end, decoders separate the strings of pulses and route them to the respective receivers, thus preventing individual conversations becoming jumbled.

Electronic calculators, digital computers and a host of other electronic equipment and components respond to and process two-state signals. Individually the pulses simply say ON or OFF; collectively they are arranged and sequenced at high speed to perform the most complex of operations. They enable us to solve long and complex mathematical calculations in minutes which would otherwise take years to complete. In addition, flight path control for busy airports, flight simulation for training pilots, the automation of industrial processes, monitoring the stocking levels of a huge chain of supermarkets, accurate timing in the digital watch and so on, all rely on digital electronic devices.

The four arithmetic functions (+ − × ÷) can be performed electronically at high speed with two-state signals. We are familiar with counting in **denary**, or to the base of ten. If only two states are permitted, 0 and 1, where 0 represents the absence of a pulse and 1 the presence of a pulse, a binary code must be used to represent the numbers greater than two (Table 8.1).

A special form of mathematics to manipulate two-state functions was developed early in the nineteenth century to express logical reasoning mathematically. This mathematics, called **Boolean algebra**, after Dr Boole who was one of the first to develop it, can be applied to two-state electronic devices (often called **logic circuits**).

The two outstanding advantages of digital signals are as follows. First, electronic circuits and components can be designed to operate unambiguously in two-states (i.e. a clear distinction between the states) and to switch between states at a very high speed. Second, the transmission of messages and information can be very fast. Fig. 8.1 compares the electrical

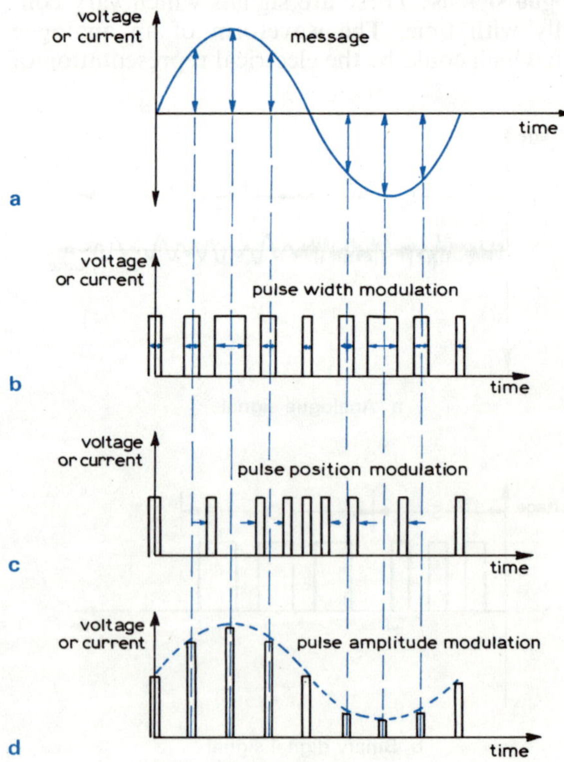

Fig. 8.2 Pulse modulation

Table 8.1 Denary and binary equivalents

Denary			Binary				Example
hundreds	tens	units	$2^3 = 8$	$2^2 = 4$	$2^1 = 2$	$2^0 = 1$	
0	0	0	0	0	0	0	
0	0	1	0	0	0	1	
0	0	2	0	0	1	0	
0	0	3	0	0	1	1	$\Rightarrow 2^0 + 2^1 = 3$
0	0	4	0	1	0	0	
0	0	5	0	1	0	1	
0	0	6	0	1	1	0	
0	0	7	0	1	1	1	
0	0	8	1	0	0	0	
0	0	9	1	0	0	1	$\Rightarrow 2^3 + 2^0 = 9$
0	1	0	1	0	1	0	
0	1	1	1	0	1	1	
0	1	2	1	1	0	0	
0	1	3	1	1	0	1	
0	1	4	1	1	1	0	$\Rightarrow 2^3 + 2^2 + 2^1 = 14$
0	1	5	1	1	1	1	

analogue of someone saying 'so' with a digital equivalent. The time duration of the analogue signal is approximately one second; the digitised code could be transmitted in less than a microsecond. This means that the majority of the words in the English language (about 490,000 words) could be transmitted in less than a second!

A further advantage is that the digital code has greater immunity to electrical distortion known as noise (for a full discussion of noise, see Chapter 11). With the addition of electrical noise an analogue signal may become unrecognisable, but the digital message still can be understood as long as the presence or absence of a pulse can be detected.

8.3 Logic levels

To represent the binary digits of 1 and 0 in digital electronics, one of two conventions can be employed; positive logic, or negative logic.

Positive logic: This has the logic 1 state represented by a more positive voltage than the logic 0 level (Fig. 8.3a). As precise voltages are difficult to ensure because of losses and component tolerances, a range of voltages is defined for each state (e.g. logic 0: 0 V to 0.8 V and logic 1: 2.4 V to 5 V).

Negative logic: This defines the logic 1 voltage level as negative with respect to the logic 0 level (Fig. 8.3b).

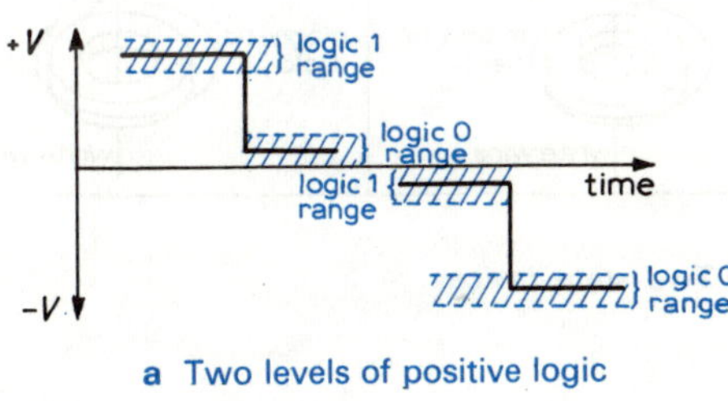

a Two levels of positive logic

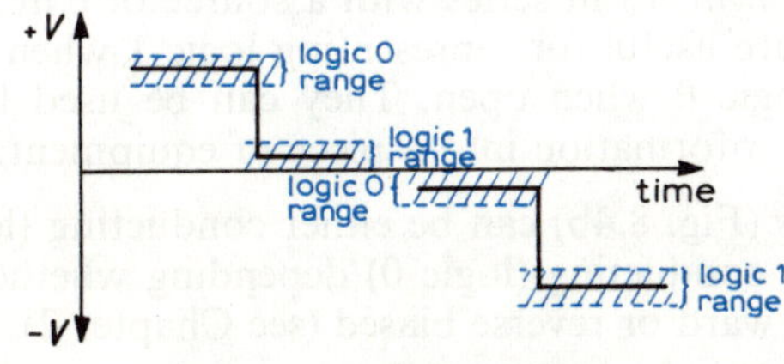

b Two levels of negative logic

Fig. 8.3 Logic levels

8.4 Two-state devices

Devices which represent or respond to digital signals are called **two-state devices**; the states defined as logic 1 and logic 0. A selection of such devices is listed below and illustrated in Fig. 8.4.

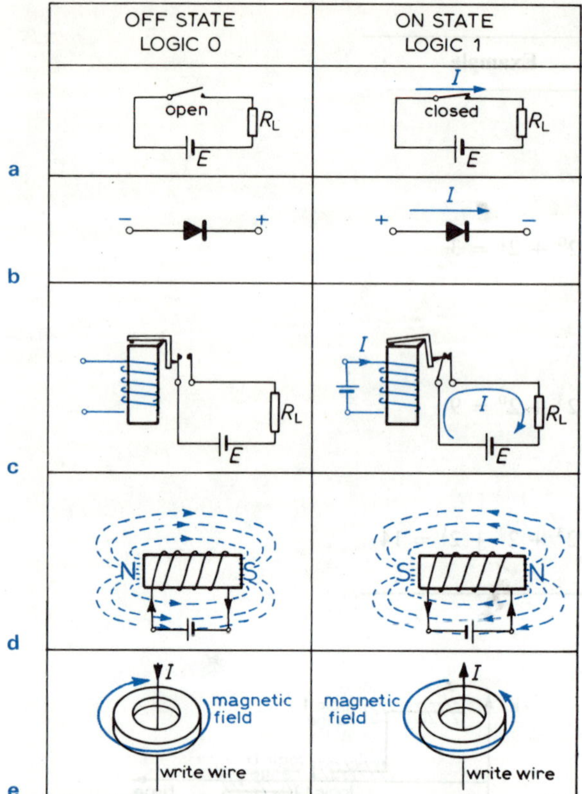

Fig. 8.4 Two-state devices

Make switches in series with a source of e.m.f. (Fig. 8.4a) are useful for representing logic 1 when closed and logic 0 when open. They can be used to feed digital information into a piece of equipment.

Diodes (Fig. 8.4b) can be either conducting (logic 1) or not conducting (logic 0) depending whether they are forward or reverse biased (see Chapter 2).

The contacts of a *relay* (Fig. 8.4c) can be open when the coil is unenergised, closed when energised.

The magnetic field of an *electro-magnet* (Fig. 8.4d) can be energised in one of two ways depending on the direction of current flowing through the coil.

Small *ferrite rings* can retain their magnetism (even when the power is switched off) in one of two states depending on the direction of current pulsed through a wire (called a write wire). Arrays of these ferrite cores were used extensively as computer memories, but now they have been replaced largely by integrated circuit memories.

Review

Check your knowledge and understanding. You should:

a Understand the difference between analogue and digital signals.
b Know the simple examples of two state devices.
c Appreciate the advantages of using digital signals for processing and conveying information.
d Know the simple examples of information which can be communicated by two-state devices.

Exercise

8.1 Identify which two of the following list are the advantages of two-state communication;
a the two states are easily identified
b simpler equipment required
c two state devices are cheaper than others
d better noise immunity

8.2 Compile a list of examples of two-state devices and methods of signalling (electronic and non-electronic).

8.3 The following incomplete table refers to the conditions of the lamp circuit below. Draw the table on a sheet of paper then complete it.

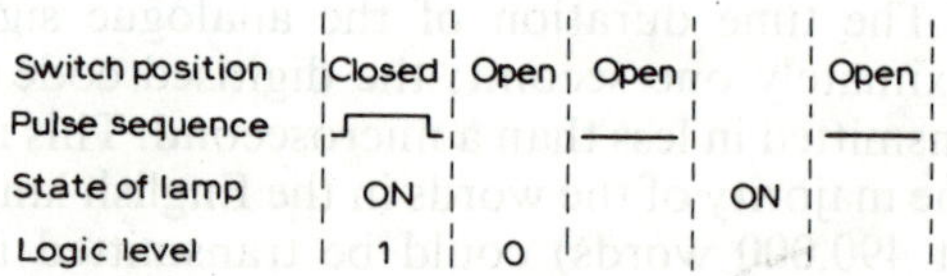

Switch position	Closed	Open	Open		Open
Pulse sequence					
State of lamp	ON			ON	
Logic level	1	0			

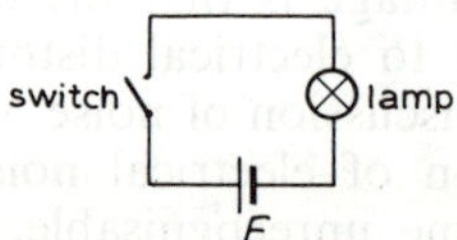

8.5 Logic gates

Special electronic circuits called **logic gates** have been developed to handle digital signals. They are so-called because they open and close electronic paths for digital signals.

There are three basic logic gates; the **AND gate**, the **OR gate**, and the **NOT gate**. (Using combinations of these, two more can be made, the **NAND** and the **NOR**, although we are not primarily concerned with these two in this text.)

At the heart of the most complicated digital equipment can be found arrangements of these basic logic gates.

To understand AND and OR gates it helps to look at a simple example outside electronics. Consider a door with two bolts, one labelled *A* the other *B*. To open the door requires the releasing of both bolts *A* AND *B*, which is an AND function. For the door to remain locked requires only one bolt to be fastened, *A* OR *B*, which is an OR function.

Electronic gates have input ports and output ports (Fig. 8.5a). Consider gates of two inputs labelled *A* and *B* and one output labelled *O*/*P*. The logic level present at the output depends on the state of the logic levels at the two inputs at a given instant in time.

Function of the AND gate

The British Standard circuit symbol of the AND gate is shown in Fig. 8.5b. Other AND symbols which may be met are shown in Fig. 8.6a. The table in Fig. 8.5b describes the output logic states of the two-input AND gate for given input conditions. A logic 1 level presented to both inputs *A* AND *B* is the only condition to result in an output of logic 1. For all other conditions, the output is logic 0. Tables showing input and output conditions are called **truth tables.**

Instead of writing the word AND when describing this function, the **dot symbol** can be used;

$$A \,.\, B \text{ means } A \text{ AND } B.$$

This a **Boolean expression** as used in Boolean algebraic relationships.

A simple practical application of the AND gate is a machine safety device. Before a dangerous machine can be switched on, let us suppose that two conditions have to be met. One condition is that the power has to be connected, the other is that a safety guard has to be in position which operates a micro-switch. If the connection of the power is represented by logic signal *A* and the closing of the safety guard by logic signal *B*, the Boolean expression is

$$A \,.\, B = 1$$

where 1 is the output signal which enables the machine to be started. AND gates can have more than two inputs. The symbol for the three input AND gate and its truth table are shown in Fig. 8.10a. To output logic 1, all three inputs have to be at logic 1. All other combinations at the inputs give a logic 0 output.

Therefore $output = A \,.\, B \,.\, C$

Function of the OR gate

Fig. 8.5c shows the truth table and British Standard circuit symbol for a two-input OR gate. Other OR symbols which may be met are shown in Fig. 8.6b.

To output a logic 1 the OR gate requires that either one or both inputs are held at logic 1.

a general logic gate

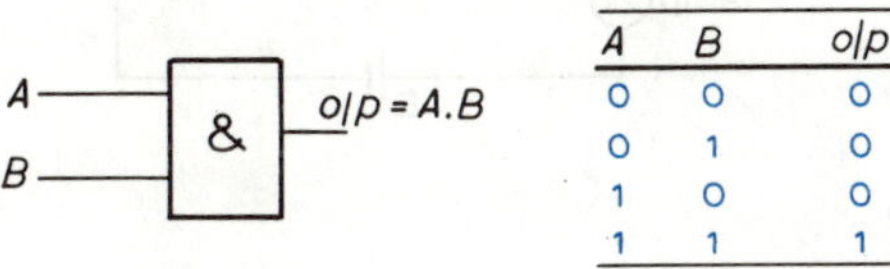

A	B	o/p
0	0	0
0	1	0
1	0	0
1	1	1

b 2-input AND gate

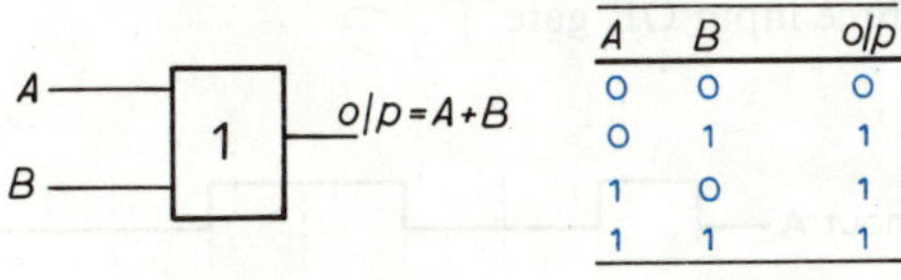

A	B	o/p
0	0	0
0	1	1
1	0	1
1	1	1

c 2-input OR gate

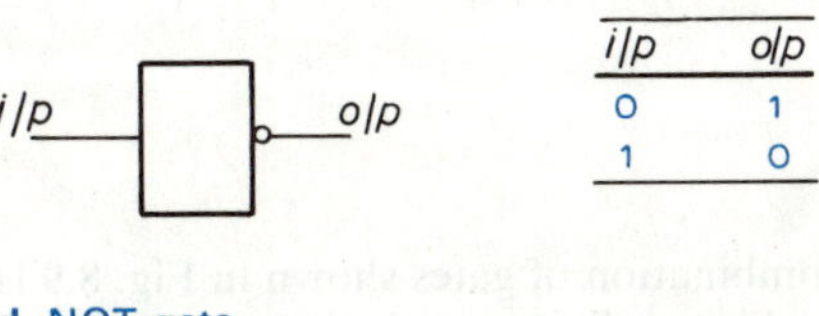

i/p	o/p
0	1
1	0

d NOT gate

Fig. 8.5 Current BS logic symbols

a AND

b OR

c NOT

Fig. 8.6 Gate symbols other than current BS

A simple practical example of the use of the OR gate is in the control of a pelican crossing. To change the traffic lights to red, a button either side of the road can be pressed. If one button initiates a pulse on input A and the other a pulse on input B, the condition A OR B is required for an output of logic 1 to the lights control circuitry. The Boolean algebraic symbol for the OR function is +.

Hence $A + B$ means A OR B.

The three input OR gate (Fig. 8.10b) requires logic 1 at any one, any two or all three inputs to produce a logic 1 at the output. In Boolean form

$$output = A + B + C$$

Function of the NOT gate

The NOT gate (Fig. 8.5d) has only one input and one output. Its function is to interchange logic 1 for logic 0 and vice versa; the output is the inverse of the input. The NOT gate is sometimes called an **inverter**. Other NOT symbols which may be met are shown in Fig. 8.6c.

The Boolean symbol for the NOT function is to place a bar over the letter designation:

$\bar{A}$ means NOT A.

If a NOT gate is placed after an AND gate, the combination provides a NOT AND or NAND function. The NAND truth table shows the output of the AND gate inverted. Likewise a NOR function can be made from NOT OR.

Review

Check your knowledge and understanding. You should:

- **a** Be able to state the logical functions of the AND, OR, and NOT gates.
- **b** Know how to construct truth tables for three input AND and OR gates and for a NOT gate.
- **c** Know the Boolean symbols for AND, OR and NOT.
- **d** Know the BS circuit symbols for the AND, OR and NOT gates.
- **e** Recognise the superseded BS symbols for the AND, OR and NOT gates.

Exercise

8.4 Identify the basic logic element which satisfies
- **a** Output 1 of the truth table below
- **b** Output 2 of the truth table below.

A	B	C	$O/P1$	$O/P2$
0	0	0	0	0
0	0	1	1	0
0	1	0	1	0
0	1	1	1	0
1	0	0	1	0
1	0	1	1	0
1	1	0	1	0
1	1	1	1	1

8.5 Construct the truth table including all the switch combinations in the circuit of Fig. 8.7, labelling the switch states as OPEN and CLOSED and the lamp states as ON or OFF

e.g. | A | B | C | LAMP |
|---|---|---|---|
| Open | Open | Open | OFF |

etc. Identify which logic function this circuit represents.

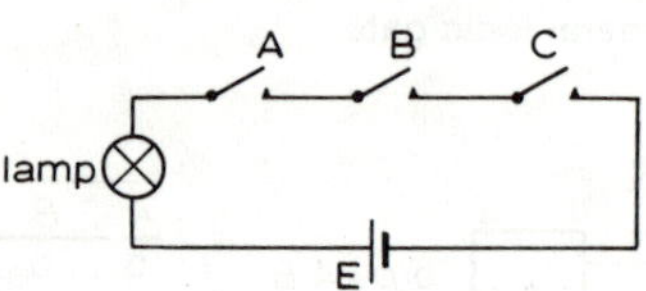

Fig. 8.7

8.6 Draw the corresponding output waveforms if the waveforms of Fig. 8.8 are simultaneously applied to the inputs of
- **a** a three input AND gate
- **b** a three input OR gate

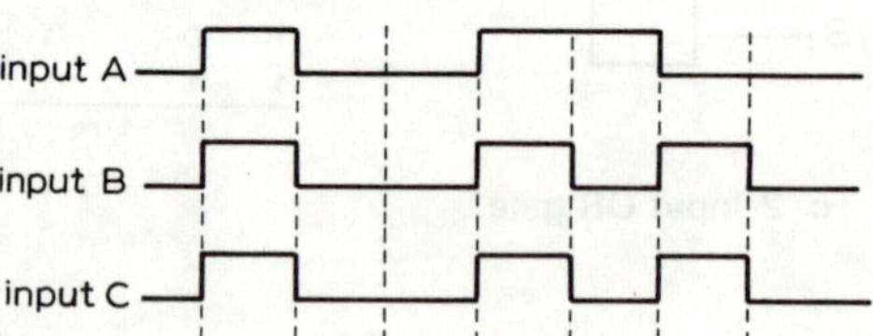

Fig. 8.8

8.7 The combination of gates shown in Fig. 8.9 is designed to add two binary digits, producing a sum S and carry C, this is called a half-adder which is used in computers and calculators. On a separate sheet of paper complete the truth

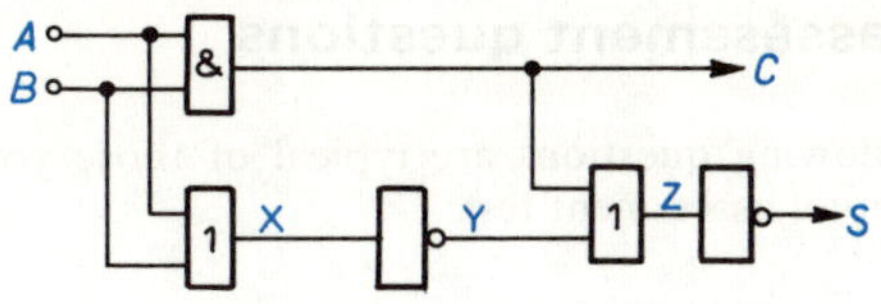

Fig. 8.9

table showing the logic states at points X, Y and Z and the two outputs C and S.

A	B	X	Y	Z	S	C
0	0					
0	1					
1	0					
1	1					

8.6 Simple electronic logic circuits

Examples of **diode-resistor logic** (d.r.l.) gates are shown in Fig. 8.10.

When any of the inputs, A, B or C, of the AND gate (Fig. 8.10) is at zero volts (logic 0), the respective diode conducts and current flows through R_L. Because only a small barrier voltage V_B is developed across a conducting diode (typically 0.2 V for germanium, 0.6 V for silicon), the output is V_B above zero, i.e. low enough for the output to be considered logic 0. Only when all the inputs are at logic 1 does no current flow. As there is no potential difference across R_L in this situation, the voltage is the same both sides. The output is therefore logic 1.

For the OR gate (Fig. 8.10b), whenever logic 1 is present at input A, B or C, the respective diode conducts. The output then is slightly less than the diode forward logic 1 level because of the forward voltage drop across the diode.

8.7 Transistor as a switch

Bipolar transistors can be used as a two-state devices. They can be switched ON or OFF depending on the input conditions. We can imagine the transistor as a switch with contacts between the collector and the emitter (Fig. 8.11a). The closing of these imaginary contacts is controlled by a current in the base, enabling current to flow from the collector to the emitter.

For the purpose of amplifying a signal using the transistor, a base bias current is provided by R_B (see Chapter 4). In a switching application, no such biasing is required (Fig. 8.11b).

Throughout the following description positive logic is assumed. When the input at the base is low

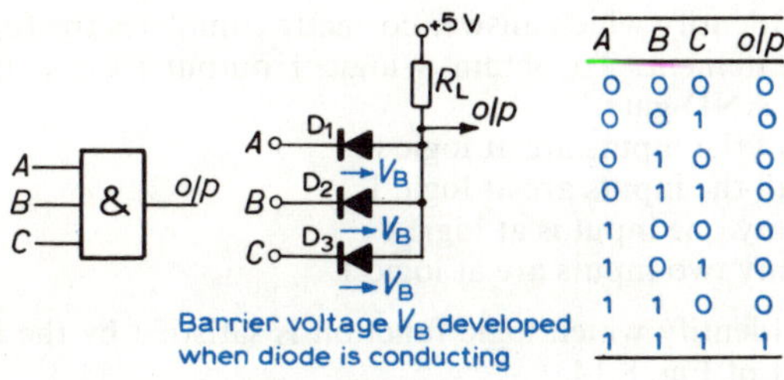

A	B	C	o/p
0	0	0	0
0	0	1	0
0	1	0	0
0	1	1	0
1	0	0	0
1	0	1	0
1	1	0	0
1	1	1	1

a three input AND gate

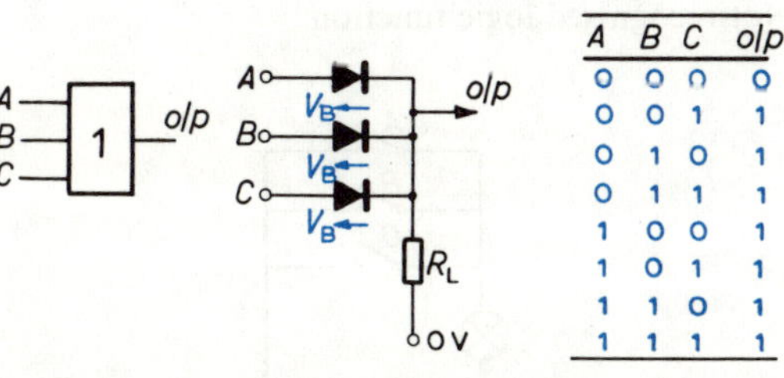

A	B	C	o/p
0	0	0	0
0	0	1	1
0	1	0	1
0	1	1	1
1	0	0	1
1	0	1	1
1	1	0	1
1	1	1	1

b three input OR gate

Fig. 8.10

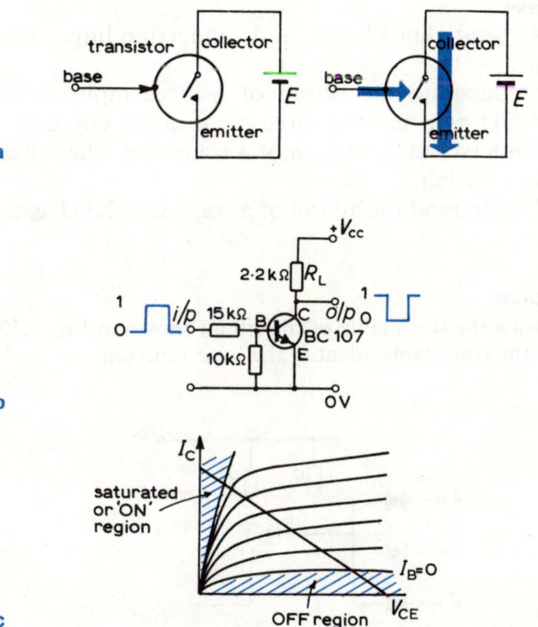

Fig. 8.11 The transistor as a switch

(logic 0), the output at the collector is high (logic 1). This is so because when no current flows in the base, no current flows in the collector and the output voltage V_{CE} is high (remember $V_{CE} = V_{CC} - I_C R_L$, Eqn. 4.7. When $I_C = 0$, $V_{CE} = V_{CC}$). The transistor is held in the OFF region of the output characteristics (Fig. 8.11c).

Conversely, when the input is high (logic 1), I_B gives rise to I_C because of the transistor's amplifying ability. The transistor is switched to the ON region of the output characteristics. As can be seen from Fig. 8.11c, the value of V_{CE} falls to logic 0 level when the transistor is saturated.

Because of the circuit's 180° degree phase shift, the transistor switch can be used as a NOT gate or inverter. By providing multiple inputs to the transistor with arrangements of resistors (**resistor-transistor logic, r.t.l.**), diodes (**diode-transistor logic, d.t.l.**) or other transistors (**transistor-transistor logic, t.t.l.**) all the basic logic functions can be produced.

The main advantage of transistor gates is that each gate gives a true logic 1 output (always just less than V_{CC}). This allows many transistor gates to be **cascaded** (connected in series) with no reduction of the logic 1 level. The number of diode gates cascaded is limited by the successive voltage drops across the diodes, which could bring down the logic 1 level close to that of logic 'O'.

Review

Check your knowledge and understanding. You should:

- **a** Understand the action of a three input diode AND gate and the three input diode OR gate.
- **b** Understand the action of a transistor when used as a switch.
- **c** Understand the action of a transistor NOT gate.

Exercise

8.8 Draw the truth table of the circuit shown in Fig. 8.12. From the truth table, identify the logic function.

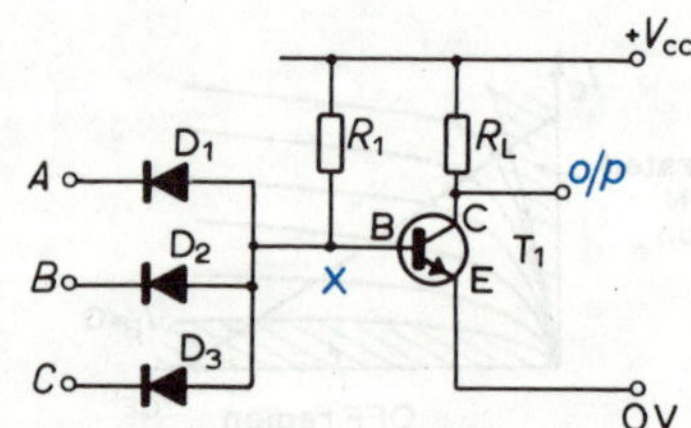

Fig. 8.12

Self assessment questions

The following questions are typical of those you will be asked in an assessment test:

Multiple choice questions

8.9 Identify which of the following give rise to two-state signals;

- **a** railway signal
- **b** digital voltmeter
- **c** motorway hazard warning lights (showing lane closures etc.)
- **d** ambulance emergency light
- **e** windvane
- **f** power indicator lamp on a piece of electrical equipment.

8.10 Identify which of the symbols shown in Fig. 8.13 are symbols for AND gates and which for OR gates. Which two are the present BS symbols?

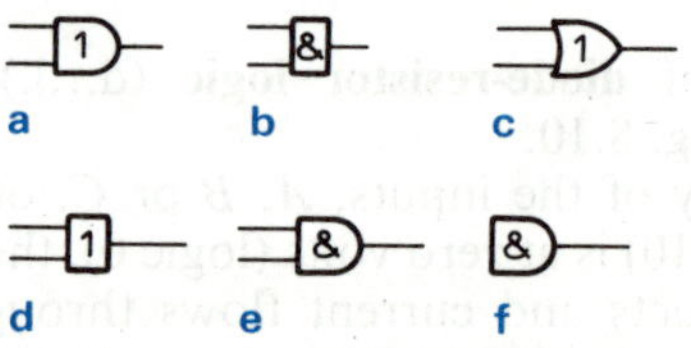

Fig. 8.13

8.11 Identify which answer correctly completes the following statement. To obtain a logic 1 output from a three-input AND gate

- **a** all the inputs are at logic 0
- **b** all the inputs are at logic 1
- **c** any one input is at logic 1
- **d** any two inputs are at logic 1

8.12 Identify which logic function is satisfied by the lamp circuit of Fig. 8.14.

- **a** AND
- **b** OR
- **c** NOT
- **d** Not recognised logic function

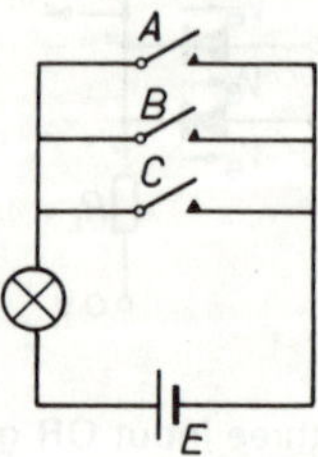

Fig. 8.14

8.13 From the truth table below identify the correct output for the logic circuit shown in Fig. 8.15.

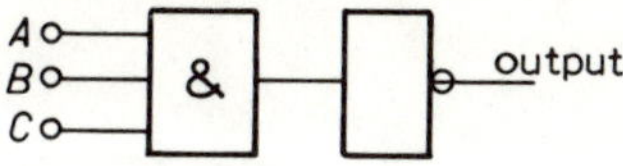

Fig. 8.15

			OUTPUT		
A	*B*	*C*	**a**	**b**	**c**
0	0	0	0	1	1
0	0	1	1	1	0
0	1	0	1	1	0
0	1	1	1	1	0
1	0	0	1	1	0
1	0	1	1	1	0
1	1	0	1	1	0
1	1	1	1	0	0

Short answer questions

8.14 Briefly describe methods of communicating information by two-state signals.

8.15 Describe four simple examples of two-state electrical devices, indicating their logic gates.

8.16 Describe the logical function of AND, OR and NOT. Draw their present BS symbols and state the Boolean symbol for each.

8.17 Draw the circuit diagram and explain the action of
a a three-input diode-resistor AND gate
b a three-input diode-resistor OR gate.

8.18 A transistor can be used as a switch. What region is the transistor driven into with a logic 1 signal at the input? What is the logic output? What is the transistor at logic 0 at the input, and what is logic of the output? What gate does this perform the function of?

8.19 Describe the operation of the diode OR gate and draw the truth table of its logic function.

Part B: Electronics Level 3

Chapters 9 to 15 cover the TEC Level Three Electronics but draw heavily on the knowledge and understanding obtained by the reader from the first part of this book.

9 Field effect transistor

This chapter describes the structure, action and application of the **field effect transistor (FET)**. The FET is a semiconductor device which, like a bipolar transistor, is able to amplify and switch electrical signals. The aims of this chapter are as follows:

- to describe the construction and basic operation of the various types of FET
- to describe the parameters of FETs
- to provide an understanding of the amplifying properties of FETs
- to provide an appreciation of the differences between FETs, bipolar transistors and thermionic triodes
- to describe the precautions necessary when using FETs
- to appreciate the need for high frequency compensation when using FETs
- to provide an understanding of the switching properties of FETs.

9.1 Introduction

The FET operates more like a thermionic diode than a bipolar transistor, its input being a voltage control signal rather than a current. It has certain advantages over the bipolar transistor in certain applications; not very susceptible to temperature variations, high input resistance and particularly suitable for use in integrated circuits.

The three terminals of the FET are called, **gate** G, **source** S and **drain** D (see Fig 9.1 for the labels and symbols).

An input voltage signal to the gate establishes an electric field which controls the current, from an external source, flowing between the source and the drain. (Note the similarity between the FET and the thermionic triode.)

There are two main types of FET; the **junction gate FET (JUGFET)** and the **insulated-gate FET (IGFET)** – sometimes called a metal oxide semiconductor FET (MOSFET).

9.2 Junction-gate FET (JUGFET)

An ***n*-channel JUGFET** (Fig. 9.1b) is made from a bar of *n*-type semiconductor with heavily doped *p*-regions on two opposite sides (the symbol p^+ indicates very heavy doping densities). A ***p*-channel JUGFET** is made from a bar of *p*-type semiconductor with n^+ regions. Electrical connections are made at three points to provide the terminals for the source, drain and gate.

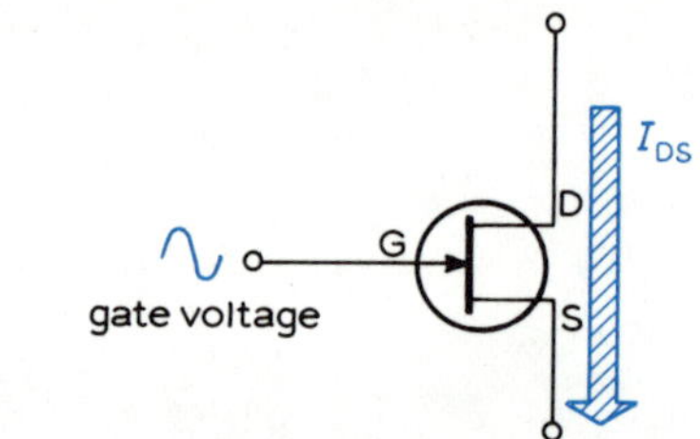

a The control of drain-source current by gate voltage

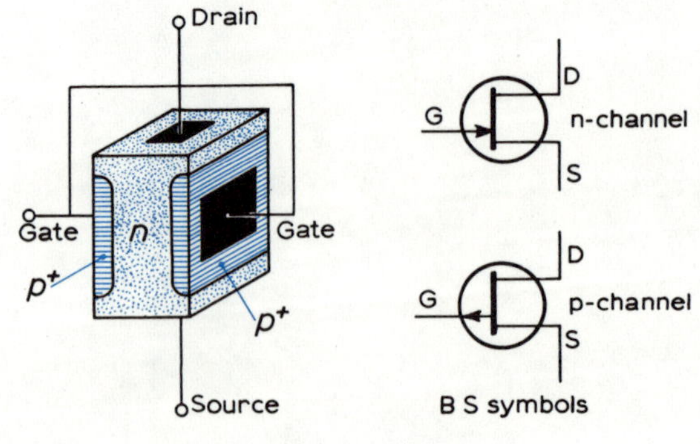

b The *n*-channel FET

Fig. 9.1 The JUGFET

The **source** S is the end into which majority carriers (electrons for *n*-type, holes for *p*-type) enter the channel.

The **drain** D is the end from which the majority carriers leave the channel.

The **gate** G comprises the two heavily doped side regions which are internally connected by a path not passing through the channel.

Depletion layers are formed at the *p-n* junctions. Because of the difference in doping densities between the regions, the depletion layer extends further into the channel than into the gate regions.

An applied reverse bias between the gate and the channel has the effect of widening the depletion layers as shown in Fig. 9.2a (remember from Section 2.3 how the potential barrier of a *p-n* junction increases and the depletion layer widens to compensate for a reverse bias p.d.). Consequently the width of the depletion layers can be controlled by the gate biasing.

As the depletion layers widen, the area of the conducting channel decreases. By constriction, current passing through the channel can be controlled (as area decreases, the resistance increases).

The shape of the depletion layer changes with the application of a p.d. along the channel (Fig. 9.2a). Because of the resistance along the channel, the drain-source voltage V_{DS} is distributed as shown by the voltage gradient in Fig. 9.2b. At the drain end the value is V_{DS}, at the source end the value is zero. As the gate potential is constant along the bar, the p.d. between the channel and the gate V_{GS} varies with the

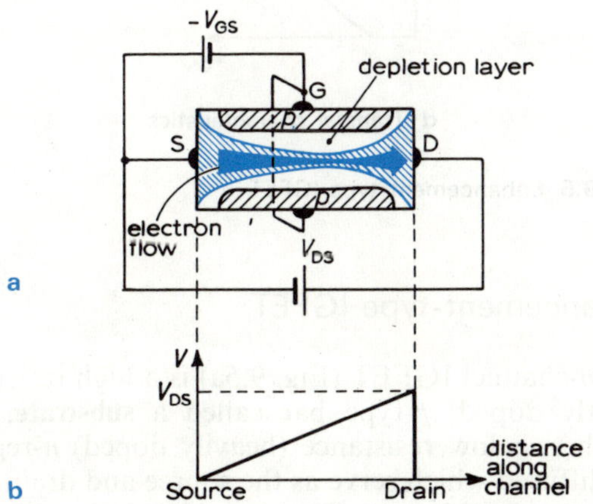

Fig. 9.2 The principle of the JUGFET
a The widening of the depletion layers
b The potential gradient along the channel

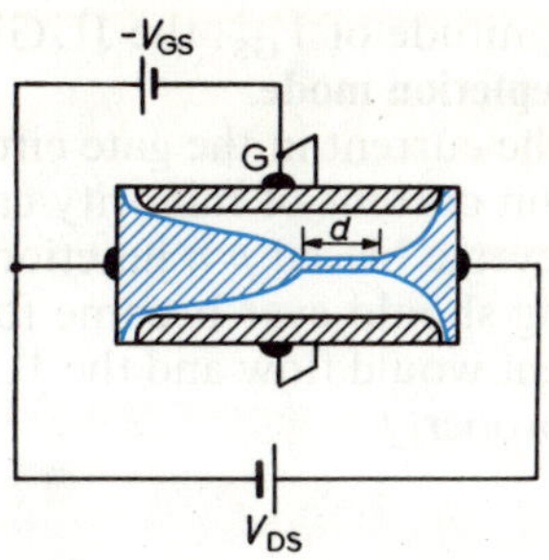

Fig. 9.3 Pinch-off

length of the channel. Therefore the depletion layer width is wider at the drain end than at the source.

The channel-gate reverse biasing ($-V_{GS}$ for *n*-channel devices, $+V_{GS}$ for *p*-channel devices) can be increased until the depletion layers almost meet across the channel. This is known as **pinch-off** (Fig. 9.3). The current through the channel I_D then becomes constant, irrespective of an increase in V_{DS} (Fig. 9.4a).

If V_{DS} is increased beyond a certain value, avalanche breakdown occurs, providing charge carriers through the normally non-conducting depletion layers. Under these circumstances the current suddenly increases which can destroy the device if maximum power ratings are exceeded.

The depletion layers do not meet at pinch-off and so do not cut-off the conducting channel altogether. Further increase in the reverse bias simply extends the length *d* of the depletion layer which is almost in contact (Fig. 9.3).

Therefore it can be deduced that there are two variables which can lead to pinch-off V_{DS} and V_{GS}; if V_{DS} is held constant, V_{GS} can be increased until pinch-off occurs, *or* if V_{GS} is held constant, V_{DS} can be increased until pinch-off occurs.

Pinch-off is usually caused by the additive effects of V_{GS} and V_{DS}.

Characteristic curves can be plotted for the JUGFET and used in the same way as those of the bipolar transistor and the thermionic valve. The information from these curves can be used to design amplifier circuits around the FET and to determine their theoretical performance (see Section 9.5).

Fig. 9.4b shows the **output** or **drain characteristics** for the *n*-channel and *p*-channel JUGFET. These show how drain current I_D varies with drain-source voltage V_{DS} for a number of constant values of V_{GS}.

The transfer characteristic (Fig. 9.4c) relates the drain current (output current) to the input voltage variations of V_{GS}.

Because the channel is depleted of carriers by an

increasing magnitude of V_{GS}, the JUGFET is said to operate in a **depletion mode**.

Normally, the current in the gate circuit is the very small saturation current of minority carriers passing through the reverse biased *p-n* junctions. If the gate-channel biasing should ever become forward biased, majority current would flow and the JUGFET would not function properly.

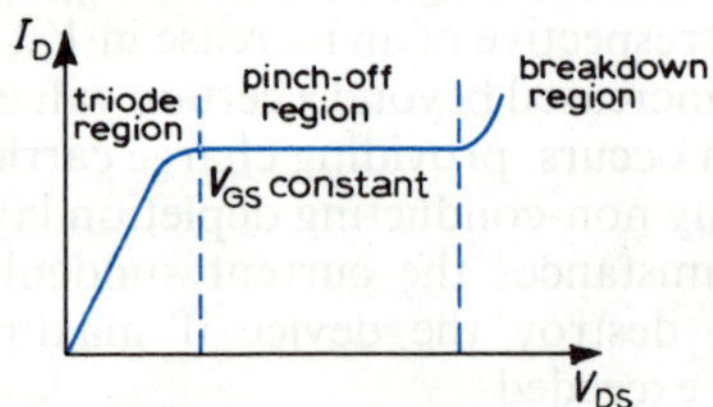

a The three regions of JUGFET drain characteristics

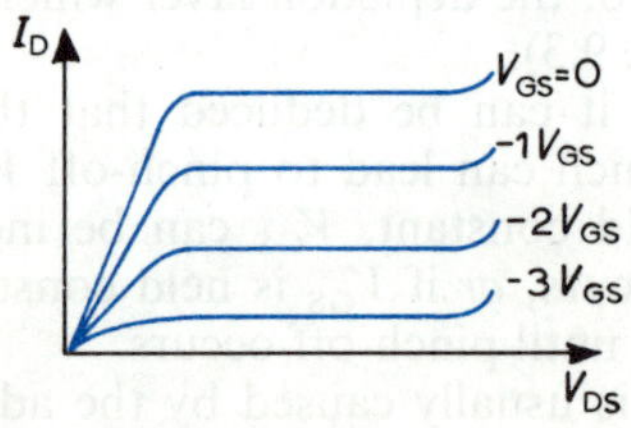

b Drain or output characteristics

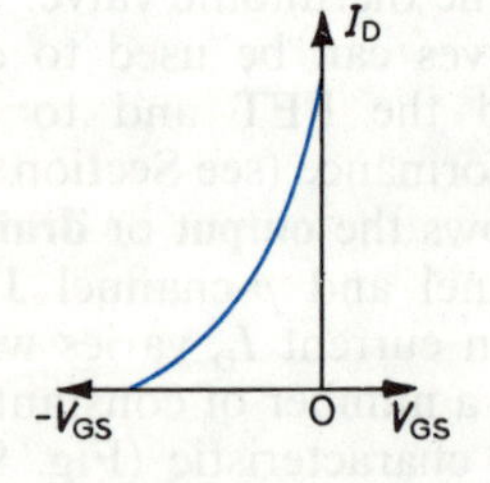

c Transfer characteristics

Fig. 9.4 JUGFET characteristics

9.3 Insulated-gate FET (IGFET)

There are two types of IGFET ; the **enhancement-type**, in which the conducting channel is enhanced or made more conductive by gate biasing, and the **diffused channel depletion-type**, in which the conducting channel is depleted of charge carriers by gate biasing.

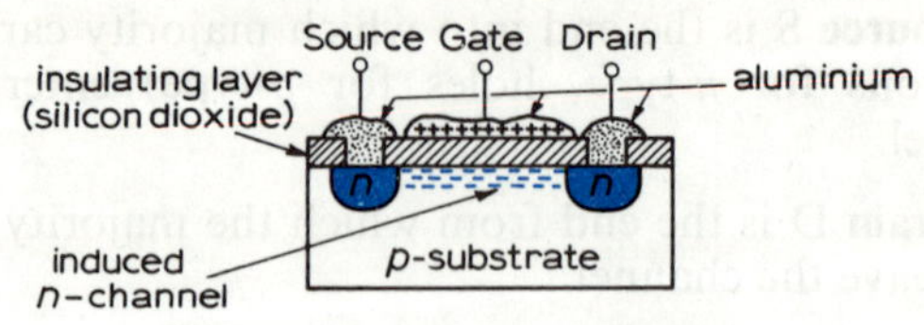

a Construction of IGFET

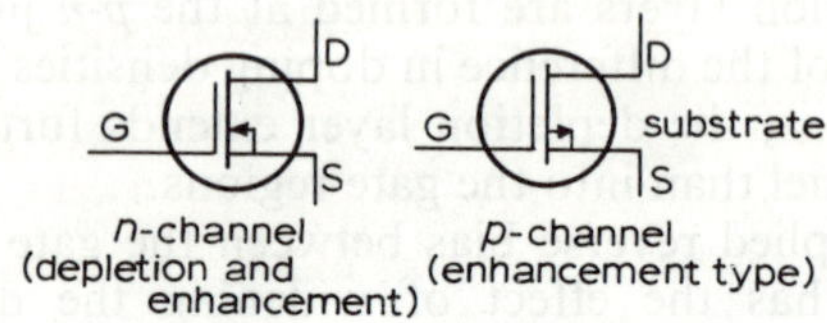

b Symbols

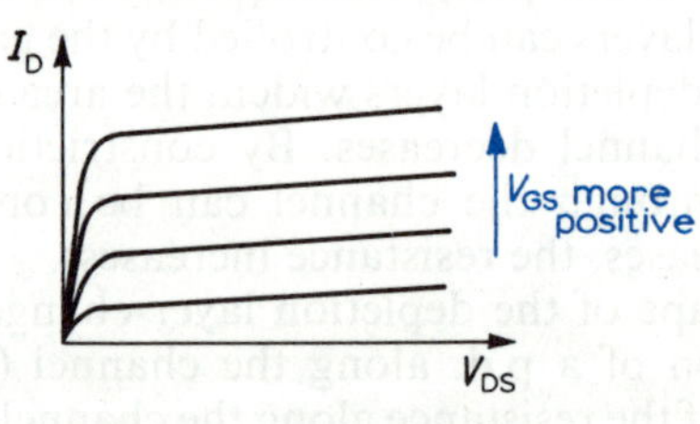

c Drain characteristics

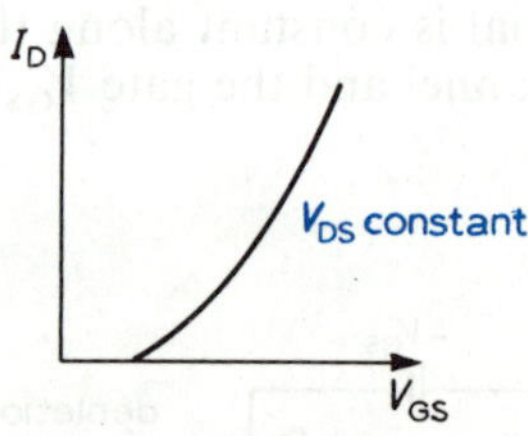

d Transfer characteristics

Fig. 9.5 Enhancement-type IGFET

Enhancement-type IGFET

The *n*-channel IGFET (Fig. 9.5a) is a high resistance (lightly doped) *p*-type bar called a **substrate**, into which two low resistance (heavily doped) *n*-regions are diffused which serve as the source and drain. The upper surface is coated with an insulating layer of silicon dioxide, leaving two small **windows** through which electrical connections are made by deposits of aluminium. An aluminium gate terminal is deposited over the channel area. Electrical contact is prevented by the layer of silicon dioxide hence the name 'insulated gate' FET. (Because the device consists of layers of **metal-oxide-semiconductor** (MOS) the IGFET is sometimes called a **MOS transistor** or **MOST**.)

With no potential at the gate, there is a very high resistance between the source and drain because there are two *p-n* junctions back to back.

The gate and *p*-substrate function as the plates of a capacitor, between which the layer of silicon dioxide serves as the dielectric. When a positive potential is applied to the gate, the capacitive effect of the device attracts electrons to a layer under the gate area. This is called an **inversion layer** or **induced *n*-channel** which serves as a conduction path between the source and drain.

As the gate potential V_{GS} is increased positively, a larger conducting channel is induced, so the drain current is greater for a given V_{DS}. Because the conduction through the channel is improved, the *n*-channel IGFET is said to operate in the enhancement mode.

The output or drain characteristics of the *n*-channel IGFET are shown in Fig. 9.5c. The transfer characteristic (I_D/V_{GS}) is shown in Fig. 9.5d. The small current flowing when $V_{GS} = 0$ is the drain-source saturation current I_{DSS}.

The gate current of the IGFET is negligible because the resistance of the insulated gate is so high (typically megohms). Even if the gate voltage became negative, no current would flow.

Note: *p*-channel IGFETs also are manufactured.

A *p*-channel IGFET is made from an *n*-substrate with *p*-type source and drain regions, a negative V_{GS} enhances the induced *p*-channel.

Diffused channel depletion-type IGFET

The only difference between the depletion-type IGFET (Fig. 9.6a) and the *n*-channel enhancement-type, is that a permanent *n*-channel is diffused between the source and drain. The electrons of the *n*-channel are repelled by the application of a negative gate potential. This creates a depletion layer (Fig. 9.6b) in this case not associated with a *p-n* junction, the thickness of which varies with V_{GS}.

Because the conducting channel is depleted by an increasingly negative V_{GS}, less drain current flows, and the device is said to work in the depletion mode.

The depletion-type IGFET shows similar characteristics to the JUGFET under these conditions. The two devices differ in behaviour when the gate potential is made positive with respect to the channel. (With the JUGFET the *p-n* junctions become forward biased and majority current flows.) With the depletion-type IGFET more negative charge is induced to the *n*-channel enhancing the conduction, i.e. I_D increases for a more positive V_{GS}.

So despite its name, the depletion-type IGFET operates in both the depletion and enhancement modes (Fig. 9.6c). This is useful when used as an

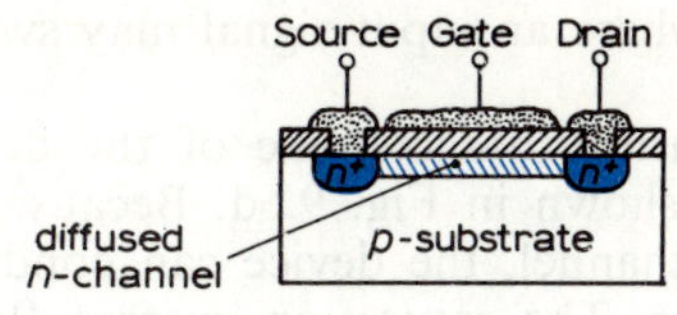

a Construction

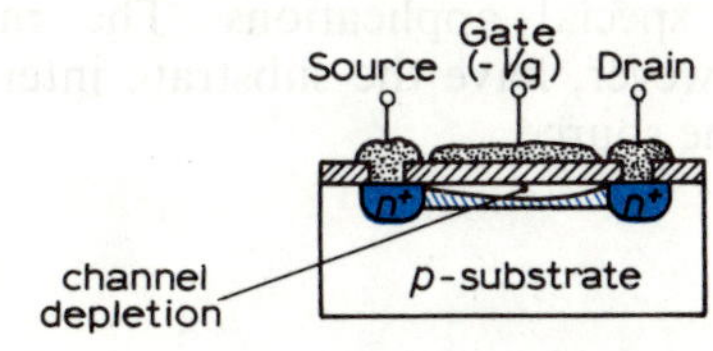

b Channel depleting

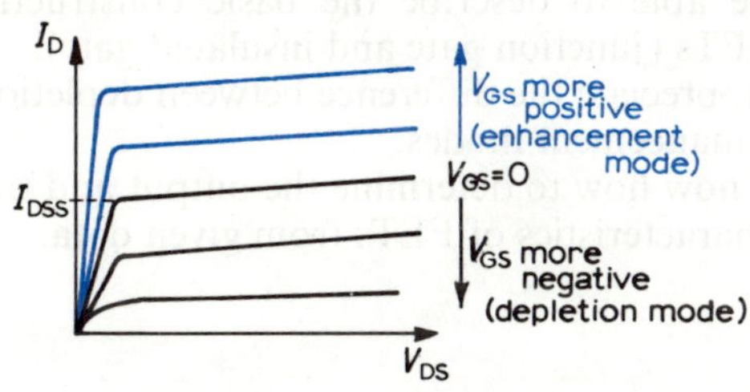

c Drain or output characteristics

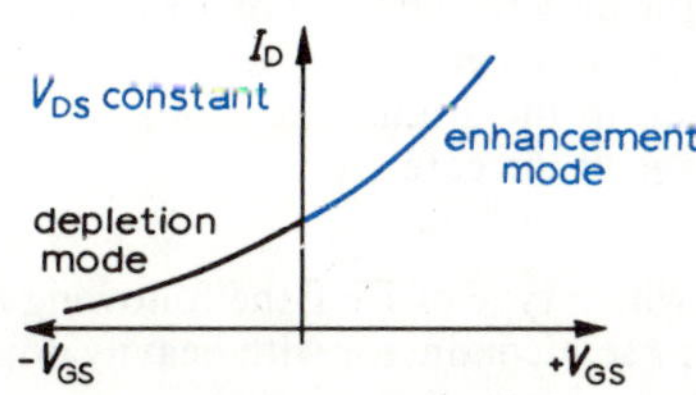

d Transfer characteristics

Fig. 9.6 Diffused channel IGFET

amplifier where an input signal may swing between polarities.

The transfer characteristic of the depletion-type IGFET is shown in Fig. 9.6d. Because of the permanent *n*-channel, the device can conduct with no gate biasing. The saturation current flowing when $V_{GS} = 0$ is marked as I_{DSS}.

Some IGFETs have a fourth terminal connected to the substrate, which can be used to vary the pinch-off voltage in special applications. The majority of devices, however, have the substrate internally connected to the source.

Review

Check your knowledge and understanding. You should:

a Be able to describe the basic construction of FETs (junction gate and insulated gate).
b Appreciate the difference between depletion and enhancement modes.
c Know how to determine the output and transfer characteristics of FETs from given data.

Exercise

9.1 Identify the answer which correctly completes the following statement.
The current through the channel of a FET is controlled by:
a the current in the gate
b the voltage at the drain
c the voltage at the gate
d the current in the source

9.2 Identify which type of FET the following describes:
A bar of *n*-type semiconductor with heavily doped *p*-regions on two opposite sides.

9.3 Identify the answer which correctly completes the following statement.
A positive gate potential applied to the *n*-channel JUGFET would:
a operate the device in the enhancement mode
b draw electrons from the channel
c operate the device in the depletion mode
d have no effect

9.4 Identify which of the following statements are true and which are false;
a the FET is a minority carrier device
b the FET is a voltage controlled device
c the JUGFET can be operated in an enhancement or depletion mode
d some IGFETs are designed to operate in enhancement and depletion modes

9.5 Identify the answer which correctly completes the following statement.
When the drain current of a JUGFET becomes constant;
a saturation has occurred
b the gate depletion layers meet
c pinch-off has occurred
d bottoming has occurred

9.6 Identify the following description as that of the enhancement or depletion mode of the FET.
The conduction through the channel is increased by gate biasing.

9.4 FET parameters

The three parameters which are used to compare the performance of different types of FET are,
mutual conductance g_m
drain resistance r_D
amplification factor μ
These are similar to those defined for the thermionic diode, Section 5.8.

From the transfer characteristics, **mutual conductance** (or **transconductance**),

$$g_m = \frac{\text{output current change}}{\text{input voltage change}} \quad \text{constant for output voltage}$$

$$= \frac{\Delta I_D}{\Delta V_{GS}}\bigg|\, V_{DS} \text{ constant} \qquad (9.1)$$

The units of mutual conductance are **siemens** S.

From the drain characteristics, **drain resistance**,

$$r_D = \frac{\text{output voltage change}}{\text{output current change}} \quad \text{constant for input voltage}$$

$$r_D = \frac{\Delta V_{DS}}{\Delta I_D}\bigg|\, V_{GS} \text{ constant} \qquad (9.2)$$

The units of drain resistance are **ohms** Ω.

Amplification factor,

$$\mu = \frac{\text{output voltage change}}{\text{input voltage change}}$$

$$\mu = \frac{\Delta V_{DS}}{\Delta V_{GS}} \qquad (9.3)$$

As with the thermionic triode, it can be shown that

$$\mu = r_D g_m \qquad (9.4)$$

9.5 FET small signal amplifiers

The FET can be connected as an amplifier in three ways;

common-source (CS)
common-drain (CD)
common-gate (CG)

However, in this text only the most usual FET amplifier connection, common-source, is considered. Typical biasing arrangements for the JUGFET and IGFET common-source amplifiers are shown in Fig. 9.7. V_{DD} is the supply voltage, (An *n*-channel FET requires a positive drain supply; a *p*-channel FET requires a negative supply.)

The *n*-channel JUGFET requires a gate bias voltage which is negative with respect to the source (Fig. 9.7a). The same condition is achieved by making the source more positive than the gate by the p.d. developed across source resistor R_S. The gate resistor R_G holds the gate at zero volts, being of a sufficiently high value to prevent the input signal being shunted to the zero volt rail.

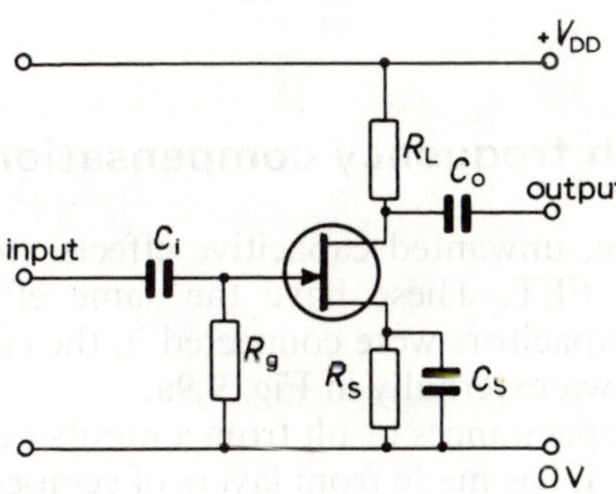

a JUGFET

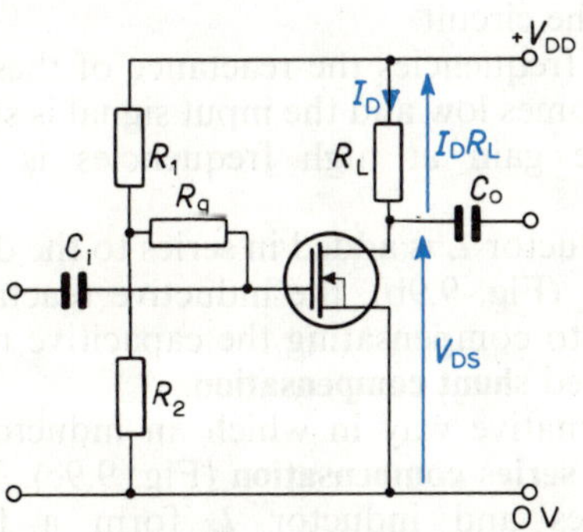

b IGFET

Fig. 9.7 Practical FET biasing circuits

The *n*-channel IGFET gate biasing (Fig. 9.7 b) is provided by the R_1/R_2 potential divider. To ensure that the input resistance is not reduced significantly by the shunting effect of R_2, a high value gate resistor R_G is included.

Similarly to the transistor and thermionic triode amplifiers, a load line can be plotted on the output characteristics and the voltage gain determined by a variation about a quiescent point Q (Fig. 9.8).

At the output of the circuit of Fig. 9.7b the voltage is

$$V_{DD} = V_{DS} + I_D R_L \qquad (9.5)$$

Rearranging

$$I_D = -\frac{V_{DS}}{R_L} + \frac{V_{DD}}{R_L} \qquad (9.6)$$

which is the FET load line equation of slope $-1/R_L$.

$$\text{When } V_{DS} = 0 \quad I_D = \frac{V_{DD}}{R_L} \qquad (9.7)$$

$$\text{When } I_D = 0 \quad V_{DS} = V_{DD} \qquad (9.8)$$

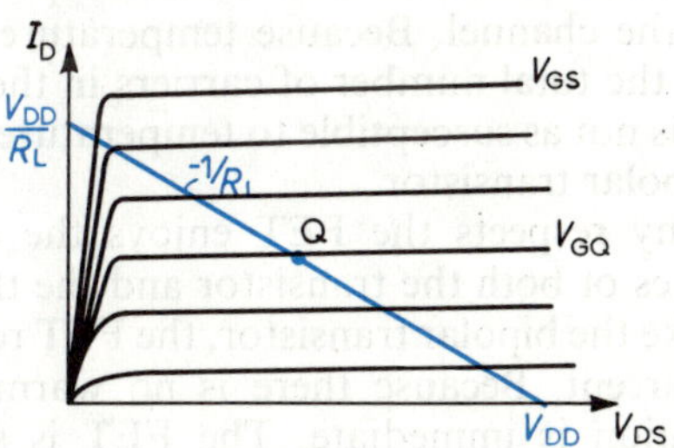

Fig. 9.8 FET load line

Eqns. 9.7 and 9.8 determine the two end points of the load line.

Knowing the input signal voltage swing about the quiescent gate voltage V_{GQ}, the stage voltage gain can be determined.

$$A_v = \frac{\Delta V_{DS}}{\Delta V_{GS}} \qquad (9.9)$$

The current gain is irrelevant as negligible current flows in the gate.

Determining the gain from the characteristics in this way is only approximate as the a.c. signal causes a departure from the variation along the d.c. load line.

The voltage gain of an FET amplifier depends largely on three factors, the amplification factor of the FET itself, the drain resistance r_D and the drain load resistance R_L. Assuming that the amplifier feeds a high resistance load which has negligible effect on the amplifier performance, an approximate expression for the gain is

$$A_v = \frac{v_o}{v_i} = \mu \frac{R_L}{R_L + r_D} \quad (9.10)$$

The derivation of this relationship is given in Appendix 3.

9.6 Comparison of the FET with the bipolar transistor and thermionic valve

Even though they are both semiconductor devices the FET and the bipolar transistor differ radically in their operation. The FET is voltage-controlled; the bipolar transistor is current-controlled (the collector current is dependent on the base current). The FET and the thermionic valve function in a similar manner although their construction is totally different.

Conduction in the FET is by one type of carrier through the channel. Because temperature has little effect on the total number of carriers in the channel, the FET is not as susceptible to temperature variation as the bipolar transistor.

In many respects the FET enjoys the combined advantages of both the transistor and the thermionic valve. Like the bipolar transistor, the FET requires no heater current. Because there is no warm-up time, its operation is immediate. The FET is small and robust.

There are similarities between the FET and the thermionic valve. Direct comparisons can be made between the electrodes. The gate and grid, source and cathode, and the drain and anode perform similar functions. Because very little current flows in the gate or grid the input resistance is very high.

The FET generates less electrical noise than the bipolar transistor or the thermionic valve and is often used at the early amplifying stage of an amplifier system.

Finally, because of their simple layered construction, FETs are ideally suited to planar technology and so are used widely in integrated circuits.

9.7 Precautions when using FETs

Owing to the FET's gate electric field it is very sensitive to any static charge which might accumulate at the gate. The IGFET can be permanently damaged, even by the charge introduced by the touch of your hand or tools such as a soldering iron or pliers.

The gate oxide of an IGFET or any MOS device is typically 10^{-7} m thick. The maximum static voltage it can withstand is typically 100 V. The human body in low humidity can generate static voltages of up to 10 000 V! Sufficient to destroy totally the oxide layer.

Once connected in a circuit, the FET is usually protected. Precautions are required when FETs are being connected into a circuit. Manufacturers provide the FETs with the terminals shortcircuited by a wire clip to ensure that all the terminals remain at the same potential. The short-circuit clip should not be removed until the device has been soldered into the circuit.

Further precautions to take are:

a Work on a metal tray which is wired to earth to prevent accumulation of charge.
b Wear an earthed metal strap around the wrist.
c Do not wear nylon clothing which readily generates static electricity.

9.8 High frequency compensation

In practice, unwanted capacitive effects are present around a FET. These have the same effect as if physical capacitors were connected in the circuit and so are shown externally in Fig. 9.9a.

These capacitances result from a number of causes. The JUGFET is made from layers of semi-conductor which act as capacitor plates. These are separated by depletion layers which act as dielectrics. The principle of operation of the IGFET is by a capacitive effect. Another cause is the stray capacitance between the wiring in the circuit.

At high frequencies the reactance of these capacitances becomes low and the input signal is shunted to earth. The gain at high frequencies is therefore reduced.

If an inductor L is added in series to the drain load resistor R_L (Fig. 9.9b), the inductive reactance goes some way to compensating the capacitive reactance. This is called **shunt compensation**.

An alternative way in which an inductor can be used is by **series compensation** (Fig. 9.9c). The stray capacitances and inductor L form a frequency sensitive network which has the effect of increasing the high frequency performance.

Sometimes both shunt and series compensation are used together.

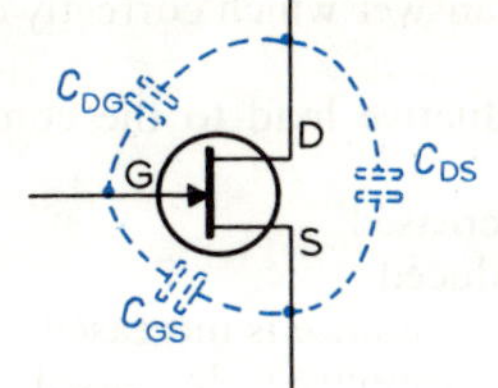

a Stray capacitances

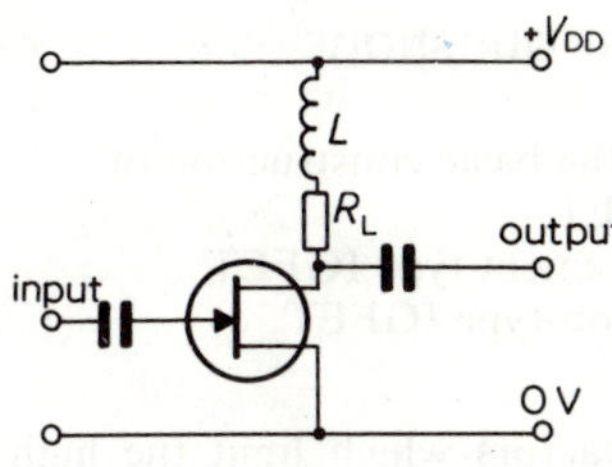

b Shunt compensation

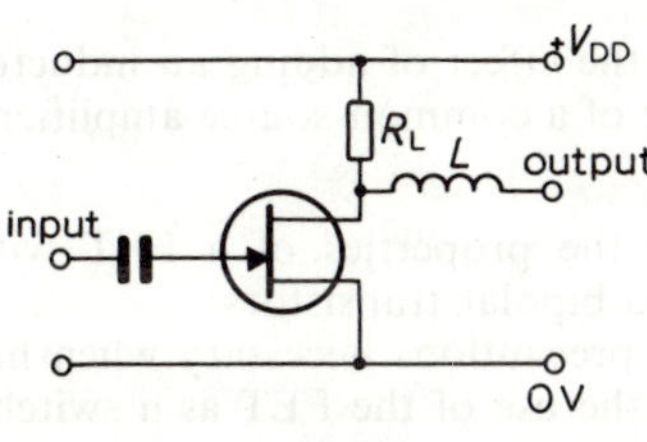

c Series compensation

Fig. 9.9 High frequency compensation

9.9 FET as a switch

Because the drain current can be turned ON and OFF under the control of a voltage at the gate, the FET can be used as a switch. The advantages of using the FET as a switch are; low power consumption (very small gate current), it can feed many subsequent FET gates because each draws little current, and it is not very susceptible to false switching brought about by random voltage fluctuations (electrical noise).

The disadvantage of the FET as a switch is that the transition between the ON and OFF states is not as rapid as bipolar because of the associated stray capacitances (see Section 9.8).

Review

Check your knowledge and understanding. You should:

a Be able to calculate the stage gain of a FET common-source amplifier stage using a resistive load.
b Appreciate the properties of a FET as compared with thermionic triodes and bipolar transistors.
c Know the precautions to be taken when using FETs.
d Understand the effect on frequency response of the FET by inductive load.
e Know the uses and advantages of the FET as a switch.

Exercise

9.7 Identify which of the following are the three parameters used to compare the performance of FETs
a drain resistance r_D
b gate conductance g_g
c amplification factor μ
d forward current gain α
e mutual conductance g_m

9.8
a Given $r_D = 400$ kΩ and $g_m = 0.45$ mS calculate the amplification factor μ.
b Given $\mu = 250$ and $r_D = 420$ kΩ calculate the mutual conductance g_m.

9.9 Calculate the voltage gain of a FET whose $r_D = 300$ kΩ, $g_m = 0.4$ mS when used in a common-source amplifier with a drain load of 25 kΩ.

9.10 The common-source drain characteristics of an *n*-channel JUGFET are given in the table below. Each characteristic is assumed linear between the two points given.
a Plot the drain characteristics and from these deduce values for the transfer characteristic for $V_{DS} = 15$ V. Plot the transfer characteristic.
b For a working point of $V_{DS} = 15$ V and $V_{GS} = -2$ V determine:
i the mutual conductance g_m
ii the drain resistance r_D
iii the amplification factor μ
c Using the values obtained in part b calculate the stage gain for a drain load resistance of 2 kΩ.

	V_{GS} (V)	0	−0.5	−1.0	−1.5	−2.0	−2.5	−3.0	−3.5	−4.0
$V_{DS} = 10$ V	I_D (mA)	9.7	8.0	6.5	5.1	4.1	3.0	1.9	1.0	0.5
$V_{DS} = 20$ V	I_D (mA)	10	8.4	6.8	5.3	4.2	3.1	2.0	1.1	0.6

Self assessment questions

The following are typical of assessment questions used to test your achievement of the objectives:

Multiple choice questions

9.11 Identify which of the following statements are true and which are false.

a The JUGFET is a majority carrier device.
b The JUGFET is a current-controlled device.
c The JUGFET is a thermionic device.
d The JUGFET can be used as an amplifier and as an electronic switch.
e The JUGFET amplifier is possible only in the common-source mode.

9.12 Identify the following description as that of a JUGFET or an IGFET:

A high resistance *p*-type bar into which two low resistance *n*-regions are diffused which serve as source and drain.

9.13 Identify which one of the following FETs conducts by an induced channel or inversion layer

a *p*-channel JUGFET
b enhancement-type IGFET
c depletion-type IGFET
d *n*-channel JUGFET

9.14 Identify the following description as that of the enhancement mode or depletion mode of FETs:
The conduction through the channel is decreased by gate biasing.

9.15 Identify the drain characteristic of a FET

a I_D plotted against V_{DS} with V_{GS} constant
b I_G plotted against V_{DS} with V_{GS} constant
c I_D plotted against V_{GS} with V_{DS} constant
d I_G plotted against V_{GS} with V_{DS} constant

9.16 Identify the transfer characteristics of a FET from the following:

a I_D plotted against V_{GS} with V_{DS} constant.
b V_{GS} plotted against V_{DS} with V_{GS} constant.
c I_G plotted against I_D with V_{DS} constant.
d V_{DS} plotted against I_G with I_G constant.

9.17 From the two lists below match the FET parameters with their correct definitions;

a mutual conductance g_m
b drain resistance r_D
c amplification factor μ

i $\dfrac{\Delta I_D}{\Delta V_{GS}} \Big|\ V_{DS}$ constant

ii $\dfrac{\Delta V_{DS}}{\Delta V_{GS}}$

iii $\dfrac{\Delta I_D}{\Delta I_G} \Big|\ V_{DS}$ constant

iv $\dfrac{\Delta V_{GS}}{\Delta I_G} \Big|\ V_{DS}$ constant

v $\dfrac{\Delta V_{DS}}{\Delta I_D} \Big|\ V_{GS}$ constant

9.18 Identify the answer which correctly completes the following statement.
By adding an inductive load to the common source amplifier

a the gain is increased
b the gain is reduced
c the frequency response is increased
d the frequency response is decreased

Short answer questions

9.19 Describe the basic construction of

a the JUGFET
b the enhancement type IGFET
c the depletion-type IGFET

9.20

a State the factors which limit the high frequency response of a common-source amplifier with a resistive load.
b Describe the effect of adding an inductor to the load resistance of a common-source amplifier.

9.21

a Compare the properties of a FET with thermionic valves and bipolar transistors.
b State the precautions necessary when handling FETs
c Describe the use of the FET as a switch and state its advantages.

9.22 Calculate the voltage gain of a FET whose $r_D = 350\ \text{k}\Omega$ and $g_m = 0.45$ mS when used in a common-source amplifier with a drain load of 20 kΩ.

9.23 The common-source drain characteristics of an *n*-channel depletion-type IGFET are given in the table below. Each characteristic is assumed constant between the two points given.

	V_{GS} (V)	0.5	0	−0.5	−1.0	−1.5
$V_{DS} =$ 5 V	I_D (mA)	15	13	11	6	4
$V_{DS} =$ 15 V	I_D (mA)	21	18	13	8	5

a Plot the output drain characteristics and from these deduce values for the transfer characteristic for $V_{DS} = 10$ V.
b For a working point of $V_{DS} = 10$ V and $V_{GS} = -1.0$ V determine from the load line the approximate small signal voltage gain when the drain load resistance is 2 KΩ and $V_{DD} = 15$ V.
c At the working point given in part (b) determine:
i mutual conductance g_m
ii drain resistance r_D
iii amplification factor μ
d Using the values obtained in part c calculate the stage gain using equation 9.10.

10 Amplifiers

This chapter describes some of the main requirements of amplifiers and how these are met. The aims of this chapter are as follows:

- to describe a typical amplifying system
- to describe how amplifiers can be specified and classified
- to provide an understanding of how the performance of amplifiers can be measured (gain, frequency response, input and output impedance, and signal amplitude limit)
- to explain some practical amplifier circuits (two-stage class A, audio-frequency power output stages, single-ended class A, class B push-pull and complementary output)
- to appreciate the disadvantage of parasitic oscillation in amplifiers and how it can be reduced.

10.1 Introduction

In Chapters 4, 5 and 9 the amplifying properties of bipolar transistors, thermionic diodes and field effect transistors were considered. Practical amplifiers usually require more than one active device plus accompanying components in order to meet a particular application. They may be high fidelity, public address, radio signal, transducer (detectors of physical change) amplifiers and so on. Electronic amplifiers make small signals larger but their variety of application means that many types of amplifier are required.

10.2 Example of an amplifying system

An amplifying system which probably is in most widespread use is the high fidelity system for amplifying radio signals and the small signals from records via a pick-up. A block diagram of such a system is shown in Fig. 10.1.

In chapter four the voltage gain of a single-stage common-source emitter amplifier was considered and found to be of the order of 10. The hi-fi system in this example could require a voltage gain of the order of 100 000 and many stages of amplification are required to achieve this. The amplifier is a **multi-stage amplifier**. The output of each stage feeds the input of the next, these stages are said to be **cascaded**.

Referring to Fig. 10.1, the radio receiver initially amplifies the **radio frequency (r.f.) signal** received from a particular transmitting station. A **detector** extracts the audio message from the r.f. which is then passed to an **audio frequency (a.f.) amplifier**. Then the signal voltage is amplified until it is strong enough to drive the **power amplifier** which feeds the loudspeaker at the output. The system differs for the playing of records only in that the signal is obtained from the pick-up on the record, pre-amplified and then fed to the a.f stage.

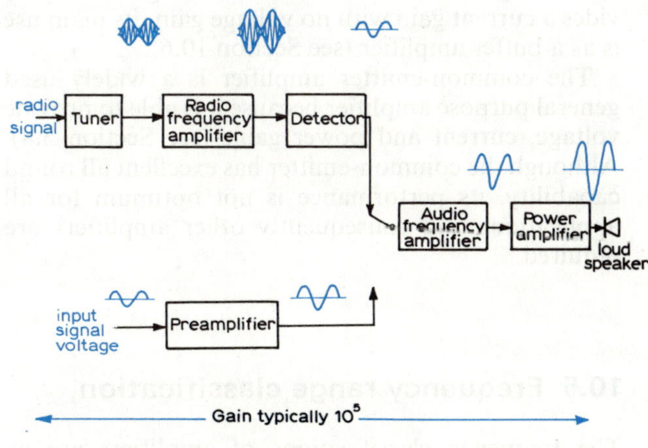

Fig. 10.1 A multi-stage amplifier system

10.3 Amplifier classification and specification

In order to describe the requirements of an electronic amplifier for an application it is necessary to specify certain parameters and its class of operation. Amplifiers can be described according to;

- **a** the electrical quantity they amplify (voltage, current and power) (see Section 10.4)
- **b** the frequency range over which they amplify (see Section 10.5)
- **c** their relative input and output impedance (buffer amplifiers) (see Section 10.6)
- **d** the method of coupling between stages (see Section 10.7)
- **e** the class of operation or how the amplifying element is biased (see Section 10.8)

10.4 Voltage, current and power amplifiers

Voltage and **current amplifiers** generally are used to increase the signal strength to a level which is sufficient to drive a **power amplifier**. Power amplifiers feed energy to the output which is a load such as a loudspeaker, electronic display, relay, transmitting aerial and so on. Power amplifiers (or large-signal amplifiers) are described further in Section 10.11.

The FET and the triode provide voltage gain because their output voltage is proportional to the gate or grid input voltages respectively. The input current is negligible. The bipolar transistor common-base configuration is also a voltage amplifier. Remember, the current gain is less than unity (see Section 3.5).

Although the common-collector configuration provides a current gain with no voltage gain, its main use is as a buffer amplifier (see Section 10.6).

The common-emitter amplifier is a widely used general purpose amplifier because it is able to provide voltage, current and power gains (see Section 3.4). Although the common-emitter has excellent all round capability, its performance is not optimum for all applications and consequently other amplifiers are required.

10.5 Frequency range classification

The frequency classifications of amplifiers are as follows:

Direct coupled amplifiers for signals which vary slowly with time.

Audio-frequency (a.f.) amplifiers for signals with a frequency range of 20 Hz to 20 kHz.

Video amplifiers for a frequency range up to several megahertz.

Radio frequency (r.f.) amplifiers for radio signals of one frequency up to hundreds of megahertz.

Ultra-high frequency amplifiers for signals of one frequency up to thousands of megahertz.

Amplifiers which simultaneously amplify a wide range of frequencies by the same amount are known as **wideband** or **untuned** amplifiers. For example, to reproduce music faithfully, an amplifier must amplify equally all frequencies in the audio-frequency (a.f.) range which is 20 Hz to 20 kHz.

If possible, the load of a wideband amplifier is kept free of reactive elements (inductance or capacitance), because if present, their response is not constant, causing the gain to vary with frequency.

Some amplifiers are designed to select a particular frequency or a narrow range of frequencies, ideally rejecting all others. These are known as **narrow band** or **tuned amplifiers**.

A tuned amplifier incorporates a **frequency selective circuit**. A typical circuit is shown in Fig. 10.2a where the load resistance of the common-emitter amplifier is replaced by the *LC* parallel combination. The output signal frequency is the resonant frequency of the tuned circuit and the **gain/frequency plot** (Fig. 10.2b) shows the narrowness of the frequency range. The tuned r.f. amplifier is said to have a **high selectivity**.

Radio frequency amplifiers are often variable-tuned so that they can pick up a range of radio stations. Variable tuning can be provided by making the inductor or the capacitor manually variable causing a change in the resonant frequency.

A desirable quality of a variable-tuned amplifier is that the selectivity remains constant over a wide range of frequencies. A variable inductor gives the most constant selectivity, but a variable capacitor is more commonly used because it is very much easier to manufacture.

10.6 Input and output impedance

The signal to an amplifier feeds into an input impedance Z_{in} (Fig. 10.3a) which is the impedance of the amplifier between the input terminals. The output, the

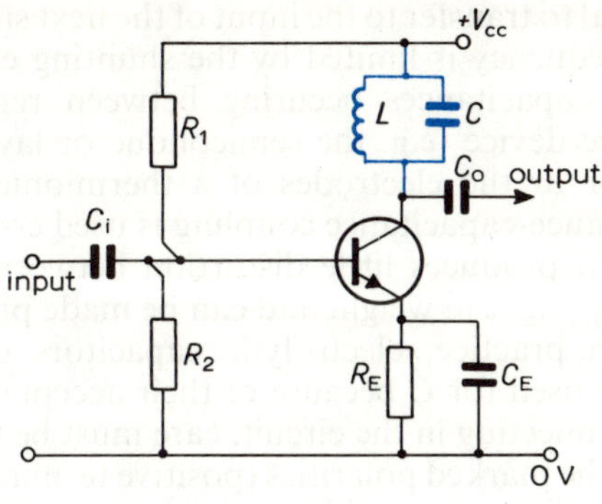

a Tuned amplifier

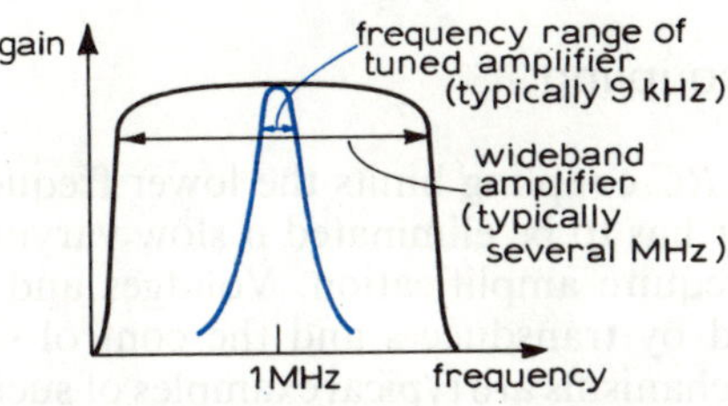

b Frequency response of tuned amplifier compared with wideband amplifier

Fig. 10.2

amplified signal, is represented by signal generator e_o which has an internal impedance Z_{out}. Z_{out} is the output impedance which is the impedance of the amplifier between the output terminals.

Fig. 10.3a shows a block diagram of a single-stage amplifier in terms of its input and output signals, v_{in} and v_o, its input and output impedances, Z_{in} and Z_{out}, and the open circuit output voltage e_o.

We refer to input and output impedance rather than resistance because reactive components are present in the amplifier. However, for the purposes of this text it is assumed that the reactances are small enough to be neglected, simplifying the impedances Z_{in} and Z_o to **input** and **output resistance**, R_{in} and R_{out}, respectively.

When amplifier stages are cascaded, it is desirable to consider that maximum signal power is transferred from one stage to the next. It can be proved mathematically that maximum power transfer occurs when the output resistance of one stage is equal to the input resistance of the next (Fig. 10.3b). Providing this situation is known as **impedance matching**.

Every signal source (microphone, pick-up, aerial, and so on) can be represented by a simple electrical circuit of an alternating e.m.f. e_s in series with its own internal resistance R_s as shown in Fig. 10.3b. Not only do successive amplifier stages need impedance matching but also matching of the internal resistance of the signal source to the amplifier, and the output stage of the amplifier to the load are required.

Circuits whose impedances are not matched are sometimes required to be connected together. For example, if a high impedance microphone is to be connected to a low input impedance amplifier. To prevent excessive loss of signal power, impedances can be matched by placing between the two circuits an intermediate stage, called a **buffer amplifier**. The prime purpose of a buffer amplifier is not to provide a signal gain, but to present to its two adjacent circuits impedances of a magnitude which prevent signal loss.

The common-collector amplifier with its high input impedance and low output impedance is often used as a buffer amplifier.

10.7 Inter-stage coupling

The passing of a signal from one stage to the next or from the amplifier to the load is done through **inter-stage coupling.**

There are three main methods of inter-stage coupling;

a resistance-capacitance coupling
b direct coupling
c transformer coupling.

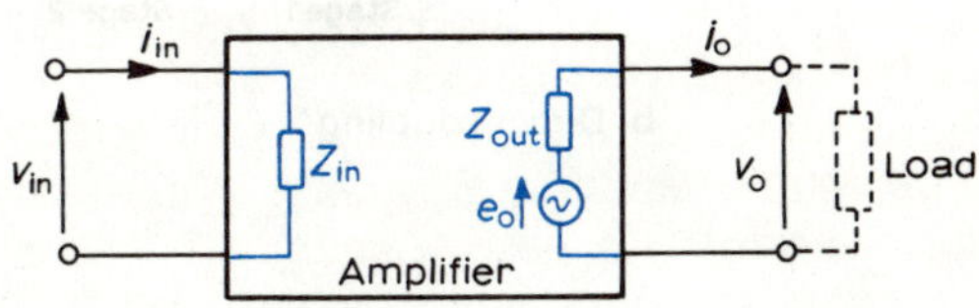

a Input and output impedance of an amplifier

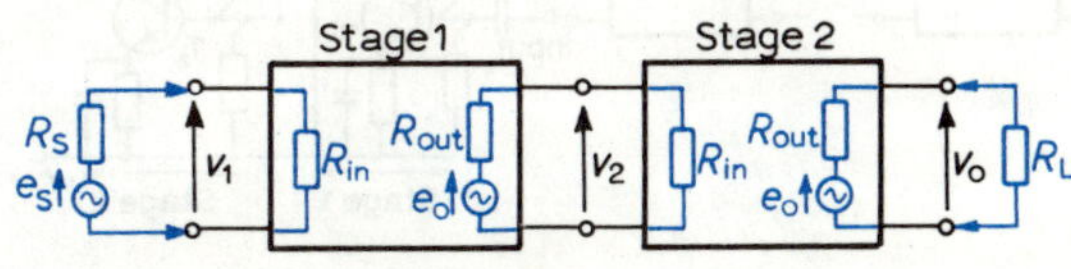

b Impedance matching

Fig. 10.3

Resistance-capacitance coupling

The use of a capacitor between the load resistor R_L of one stage and the input network of the next is the simplest method of coupling (Fig. 10.4 a). The value of C is chosen so that signal frequencies are passed with little opposition, yet the direct biasing current is blocked from passing between stages.

If the d.c. level at the collector of T_1 was connected directly to the base of T_2, T_2 would be driven into saturation by the collector current of the first. D.C. blocking is important in thermionic valve and JUGFET amplifiers because a positive potential on the grid or gate would cause an electron current to flow from the device.

At low signal frequencies, the overall gain is reduced as the reactance of C is increased allowing less signal to transfer to the input of the next stage. The upper frequency is limited by the shunting effects of internal capacitances occuring between regions of the active device (e.g. the semiconductor layers of a transistor or the electrodes of a thermionic valve).

Resistance-capacitance coupling is used commonly because it produces little distortion between stages, it is cheap, light in weight and can be made physically small. In practice, electrolytic capacitors of value 10μF are used for C because of their acceptable size. When connecting in the circuit, care must be taken to observe the marked polarities (positive terminal to the collector of *npn* since this is more positive than the base of the following stage).

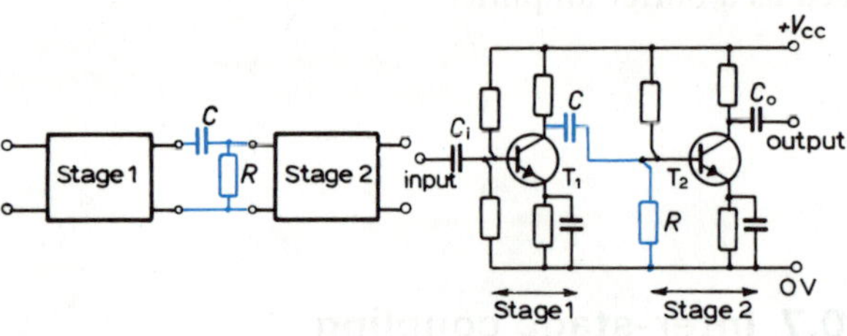

a Resistance-capacitance coupling

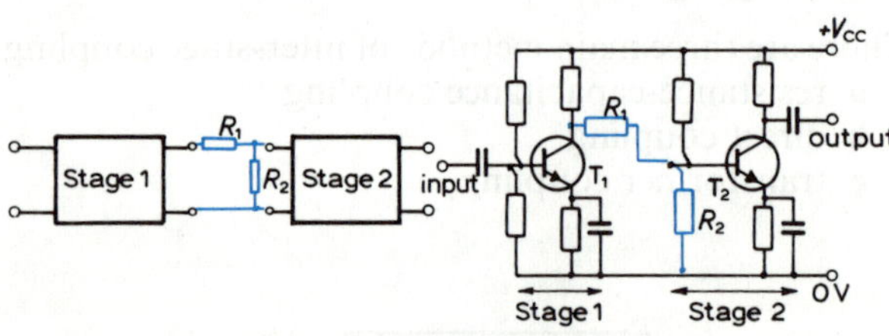

b Direct coupling

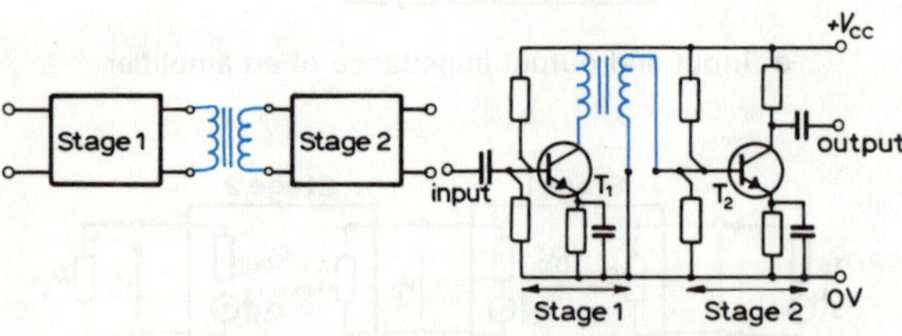

c Transformer coupling

Fig. 10.4 Inter-stage coupling

Direct coupling

Because RC coupling limits the lower frequency, the capacitor has to be eliminated if slow varying or d.c. signals require amplification. Voltages and currents produced by transducers and the control signals of servomechanisms are typical examples of such signals.

The output from one stage can be **directly coupled** to the input of the next or connected through a resistance potential divider (Fig. 10.4b). Direct coupled amplifiers require careful design to provide the correct biasing from stage to stage.

The main disadvantage of direct coupled amplifiers is that any **drift** (change in gain caused by a change in the d.c. supply, temperature or ageing of components) at an early stage is amplified by later stages.

Transformer coupling

Amplifier stages can be coupled by means of transformers (Fig. 10.4c), the primary winding normally is connected into the collector, anode or drain of one stage and the secondary winding feeds the next stage or load.

There are three advantages of transformer coupling:

a Only a.c. is passed through the transformer to the next stage or load. In some applications the d.c. supply should not reach the next stage (e.g. thermionic valve or JUGFET).

b The d.c. power loss is small because of the low resistance of the transformer primary winding.

c The primary and secondary impedances of the transformer can be chosen such that the amplifier output impedance is matched with the load. This enables maximum power gain to be realised (see Section 10.6).

Advantage can be taken of the inductive reactance of the transformer when it is required to tune an amplifier with the addition of a variable capacitor.

The main disadvantage is that the response over a wide frequency range may not be constant because of change in the primary inductive reactance and stray capacitance. There is also the possibility of non-linear magnetic characteristics in the iron core. In addition, transformers are relatively expensive components.

10.8 Class of operation

An amplifying element (bipolar transistor, FET or thermionic valve) can be biased to operate in one of four classes; class A, B, AB or C. Each class can be illustrated by its output waveform resulting from a full-wave sinusoidal input current as shown in Fig. 10.5. The class of operation is determined by the position of the quiescent point Q on the output characteristics.

Class A: Current flows in the output circuit (collector, drain or anode) for the whole cycle of the waveform (Fig. 10.5a). The Q-point is placed in the centre of the load line to accommodate the full sinewave.

Class B: Current flows in the output circuit for one half cycle of the waveform (Fig. 10.5b). The Q-point is placed at the cut-off point on the load line.

Class AB: Current flows for more than half but less than the whole cycle. The Q-point is placed on the load line somewhere between the position for classes A and B.

Class C: Current flows for less than one half cycle (Fig. 10.5c). The Q-point is placed below the cut-off point.

With class A operation up to 50% of the d.c. power supplied to the amplifier can be converted into output signal power. This is a measure of the **efficiency** of the amplifier. Actual operating efficiencies are normally lower than 25%. Because class A is normally used in low power circuits, efficiency is of secondary importance to low distortion.

Typical applications of class A amplifiers are tuned and untuned voltage amplifiers and low power audio frequency amplifiers.

The maximum theoretical efficiency of class B operation is 78.5%, although the practical efficiency is typically 50%. Because the efficiency of class B is higher than class A, it is less wasteful of power to use class B for audio frequency power amplifiers. The lost half cycle of the waveform is replaced by circuits discussed in Section 10.11.

Class C operation is often employed in oscillator and tuned radio-frequency amplifiers. Although the output waveform is severely distorted, a tuned circuit selects the required output frequency and restores the output waveform to a sinusoidal shape. Efficiencies approaching 100% are possible.

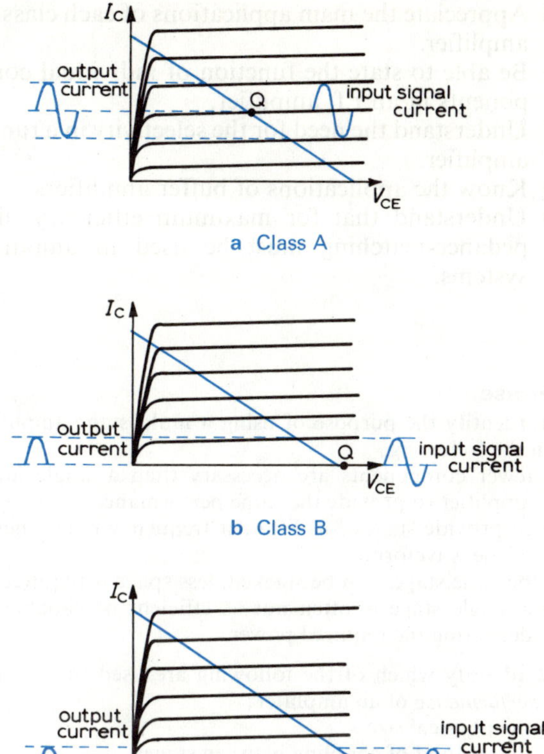

Fig. 10.5 Class of operation

Review

Check your knowledge and understanding. You should:

- **a** Be able to describe resistance-capacitance, direct and transformer inter-stage coupling of amplifiers.
- **b** Know applications of the coupling methods stated in a.
- **c** Know the biasing conditions for class A, B, AB, and C amplifier operation.

d Appreciate the main applications of each class of amplifier.
e Be able to state the function of individual components in an r.f. amplifier.
f Understand the need for the selectivity in a tuned amplifier.
g Know the applications of buffer amplifiers.
h Understand that for maximum efficiency, impedance-matching must be used in amplifier systems.

Exercise

10.1 Identify the purpose of using a multi-stage amplifier from the following;
- **a** fewer components are necessary than a single large amplifier to provide the same performance
- **b** to provide stages for different frequency components of the waveform
- **c** because stages can be stacked, less space is required
- **d** a single stage is often not as efficient or capable of delivering the required power.

10.2 Identify which of the following are used to describe the *performance* of an amplifier;
- **a** the physical size
- **b** the method of coupling between stages
- **c** the electrical quantity amplified (voltage or current)
- **d** the relative input and output impedance
- **e** components used (e.g. FET, bipolar, transistor etc)
- **f** the number of inputs and outputs
- **g** the class of operation
- **h** the frequency range amplified

10.3 Identify which of the following is normally used as a power output stage;
- **a** class C amplifier
- **b** wideband amplifier
- **c** buffer amplifier
- **d** class B amplifier

10.4 Identify which of the following statements are true and which are false;
- **a** a power amplifier delivers energy to a load
- **b** the bipolar transistor common-emitter amplifier is a voltage and current amplifier but not a power amplifier
- **c** a narrow band or tuned amplifier is used for a video signal
- **d** The coupling of stages in a multi-stage amplifier affects the gain and frequency response
- **e** The class of operation of an amplifier is fixed by the position of the quiescent point
- **f** Class C is the most efficient mode of operation.

10.9 Amplifier measurements

Amplifiers are designed to meet specifications for a particular application. However, amplifier performance is required to be measured because the practical performance generally differs slightly from the theoretical performance (factors such as variations in the components can be the cause of this). In addition, design improvements are quickly and effectively assessed by measuring amplifier performance.

In certain systems there is a requirement to connect various controls, circuits, devices and amplifiers together. It is necessary to measure the total amplifier system performance because it is not easy to obtain this from the individual specifications of the system's parts.

To evaluate performance the main measurements required are: gain, frequency response or bandwidth, input and output impedances, and signal amplitude limit.

Measurement of gain

To measure the gain of an amplifier, the circuit of Fig. 10.6 can be used. A small sinusoidal signal source is injected into the input. The output signal is developed across a dummy load R_{LL} (this is a resistor with the value of a typical true load, e.g. an 8 Ω resistor to represent a loudspeaker). The frequency of the signal generator is set to the expected mid-frequency of the amplifier (e.g. 10 kHz for an a.f. amplifier).

The input signal is measured on one input of a double-input CRO, the output signal measured on the other. (Note the connection of the earth or common leads. If these are not connected as shown, the input or output may be short-circuited to earth.) Because the input impedance of a CRO is high (typically 1 megohm) little current is drawn from the signal and the measured quantities are relatively unaffected. (High impedance a.c. voltmeters can be used in place of the CRO for these measurements.)

The voltage amplitudes of the input and output waveforms V_i and V_o are then measured.

$$\textbf{Voltage gain } A_v = \frac{V_o}{V_i} \qquad (10.1)$$

Often the power gain is required:

$$\textbf{Power gain } A_p = \frac{\text{output power}}{\text{input power}} = \frac{P_o}{P_i} \qquad (10.2)$$

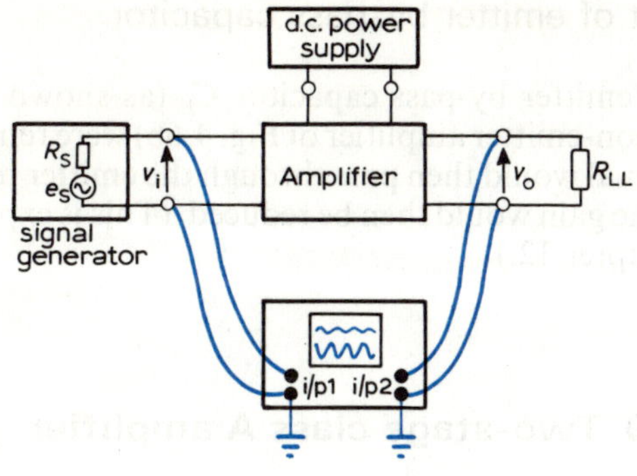

a Measurement of gain

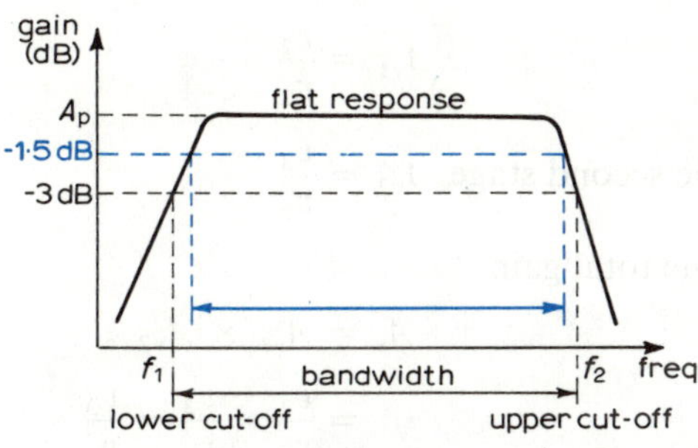

b Frequency response

Fig. 10.6 Amplifier measurements

The gain of an amplifier is often quoted as a logarithmic value with the units of the **bel**.

$$\text{Power gain } A_p = \log_{10}\left(\frac{P_o}{P_i}\right) \text{bel} \qquad (10.3)$$

The unit of the bel is often too large, so a smaller unit called the **decibel** dB is used, which is one tenth of a bel;

$$A_p = 10 \log_{10}\left(\frac{P_o}{P_i}\right) \text{dB} \qquad (10.4)$$

When a number of amplifier stages are cascaded the logarithmic gains of each stage are added, which is easier than multiplying gain ratios (see Section 10.10). This is more useful in systems where amplifiers provide gains and the inter-connecting circuitry (e.g. transmission line in a communication system) introduces **attenuation** (reduction of signal strength). A logarithmic attenuation is a negative logarithmic gain. So the individual gains and losses are respectively added and subtracted to determine the overall gain.

Assuming that the input power and output power are developed in the same magnitude of resistance, a power gain in terms of voltage gain can be deduced.

$$P_i = \frac{V_i^2}{R} \qquad (10.5)$$

$$P_o = \frac{V_o^2}{R} \qquad (10.6)$$

Substitute 10.5 and 10.6 into 10.4

$$A_p = 10 \log_{10}\left(\frac{V_o^2}{R} \times \frac{R}{V_i^2}\right)$$

$$= 10 \log_{10}\left(\frac{V_o}{V_i}\right)^2$$

$$A_p = 20 \log_{10}\left(\frac{V_o}{V_i}\right) \text{dB} \qquad (10.7)$$

Measurement of frequency response

If the gain of an amplifier is measured over a range of frequencies and a graph plotted of gain/frequency, a curve similar to Fig. 10.6b is obtained. This is a typical frequency response curve of an amplifier. Notice that the gain falls away sharply at low and high frequencies, but generally the mid-frequency gain is constant.

The useful frequency range of an amplifier is called the **bandwidth** and is determined from the frequency response curve. It is the range over which the amplifier provides an almost constant gain with no frequency distortion (not amplifying one frequency significantly more than another).

The bandwidth is measured between two frequencies known as the **low** and **high frequency cut-off** points f_1 and f_2. Cut-off is defined as the frequencies at which the power gain falls to half its mid-frequency value or by a reduction of 3 dB.

Measurement of input and output impedance

To measure the input impedance of an amplifier the relationship between the a.c. input voltage v_i and input current i_i is used;

$$Z_i = \frac{v_i}{i_i} \text{ ohm} \qquad (10.8)$$

Fig. 10.7a shows how by the inclusion of a series resistor in the input circuit, two voltmeters can be used for this measurement. The value of v_i is measured directly; the value of i_i is determined by Ohm's law from R and the voltage measured across it.

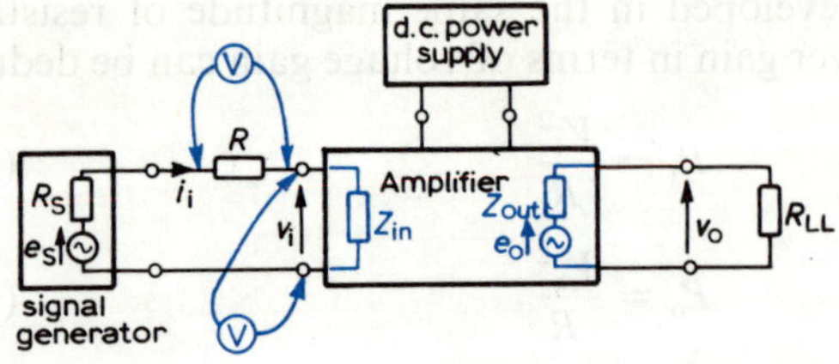

a Input impedance

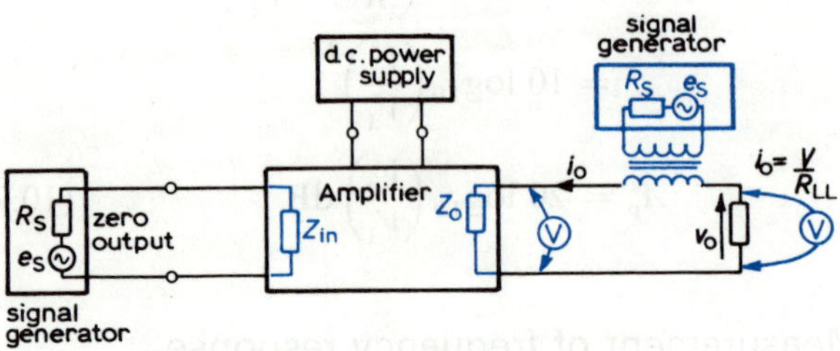

b Output impedance

Fig. 10.7 Measurement of impedance

To avoid affecting the current during measurement, a high resistance electronic voltmeter must be used. Also it must be ensured that the voltmeter terminal is not earthed, otherwise the signal is short-circuited to earth.

Measuring the output impedance requires injecting a signal into the output (Fig. 10.7b) and determining the relationship between the resulting a.c. voltage v_o and current i_o, that is

$$Z_o = \frac{v_o}{i_o} \text{ ohms} \tag{10.9}$$

The signal generator is connected through a transformer with a low turns ratio, because if connected directly it would affect the biasing conditions at the output.

The output of the signal generator at the input is set to zero but is left in the circuit because the input is required to be matched with the source, as with normal operating conditions.

Signal amplitude limit

If the amplitude of an amplifier's input signal is increased until the output waveform just distorts (clipping or bottoming as shown in Fig. 4.5) the input and output voltages can be measured as the maximum possible with that particular amplifier.

Effect of emitter by-pass capacitor

If the emitter by-pass capacitor C_E (as shown in the common-emitter amplifier of Fig. 4.6b) were removed, the signal would then pass through the emitter resistor R_E. The gain would then be reduced. (This is explained in Chapter 12.)

10.10 Two-stage class A amplifier

By means of cascading amplifier stages the overall gain becomes the product of the individual stage gains.

If the gain of the first stage is

$$A_{v1} = \frac{v_2}{v_1}$$

and the second stage $A_{v2} = \dfrac{v_3}{v_2}$

then the total gain

$$A_v = A_{v1} \times A_{v2} \tag{10.10}$$

$$= \frac{v_2}{v_1} \times \frac{v_3}{v_2} = \frac{v_3}{v_1}$$

This holds true only if interconnected stages do not affect each other. In this case the second stage must have a high input impedance and not act as a load on the first stage.

The overall bandwidth becomes narrower as the number of stages increases. With two stages, each stage has to drop only 1.5 dB for 3 dB to be dropped overall. With reference to Fig. 10.6 of Section 10.9, the bandwidth between the 1.5 db points is significantly less than between the 3 dB points of a single stage.

The overall input impedance Z_{in} is equal to the input impedance of the first stage unless feedback is applied (see Chapter 12). Likewise, the overall output impedance Z_{out} is equal to the output resistance of the final stage.

Review

Check your knowledge and understanding. You should:

a Be able to appreciate the need to measure the gain, frequency response, input and output impedances and signal amplitude of electronic amplifiers.

b Know how to measure the quantities specified in **a**.

c Understand the effect on stage gain and bandwidth of disconnecting the emitter by-pass capacitor in a common-collector amplifier.
d Be able to predict the performance of a two-stage class A amplifier.

Exercise

10.5 Identify the answer which correctly completes the following statement.
The quoted gain of an amplifier is normally;
- **a** the gain at the upper 3 dB point
- **b** maximum at mid-frequency
- **c** minimum at the lower 3 dB point
- **d** the average gain of the individual stages

10.6 Identify the answer which correctly completes the following statement.
When a number of amplifier stages are cascaded, the individual gain ratios are;
- **a** added
- **b** multiplied
- **c** subtracted
- **d** divided

10.7 Identify the answer which correctly completes the following statement.
The frequency range between the half-power points of an amplifier is called
- **a** the selectivity
- **b** the audio-frequency
- **c** the stop-band
- **d** the bandwidth

10.11 Audio-frequency power output stages

Power amplifiers are required to deliver a required amount of energy to an output device. To do this, use must be made of the maximum voltage and current swings possible without distortion, as determined by the output characteristics (Fig. 10.8). For this reason, power amplifiers are referred to as **large signal amplifiers**.

In this text only class A and B (class C introduces high distortion) audio-frequency power amplifiers are considered.

Single-ended class A power amplifiers

A bipolar common emitter class A amplifier (as shown in Fig. 10.8b) can be used as a **single-ended power amplifier**. (Single-ended means that only one output transistor is used.)

The circuit is designed around a transistor capable of providing the output power required by the load. The resistance of the output device is used as the collector load resistance in this case.

At all times during operation, the maximum power rating of the transistor $P_{(max)}$ (as quoted by the manufacturer) must not be exceeded. A curve connecting the points of maximum power can be plotted on the output characteristics as shown in Fig. 10.8a. All the power is assumed to be dissipated in the collector

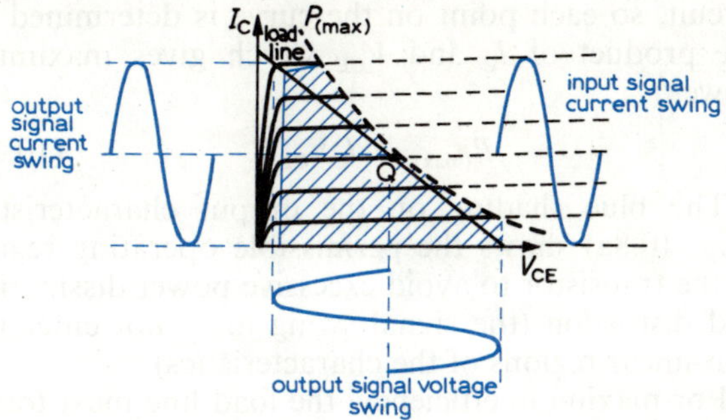

a Large signal variations

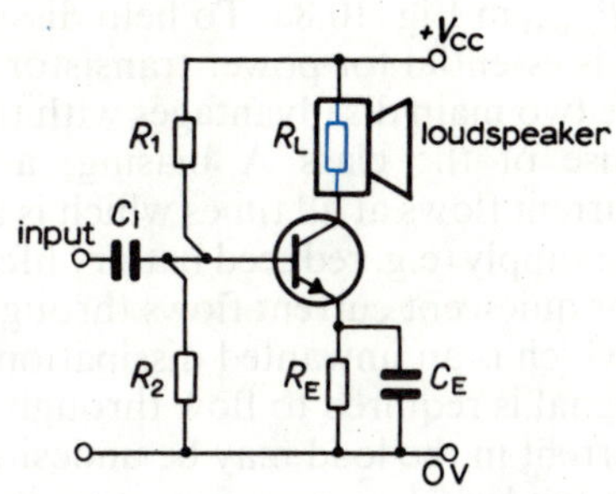

b Single-ended class A power amplifier

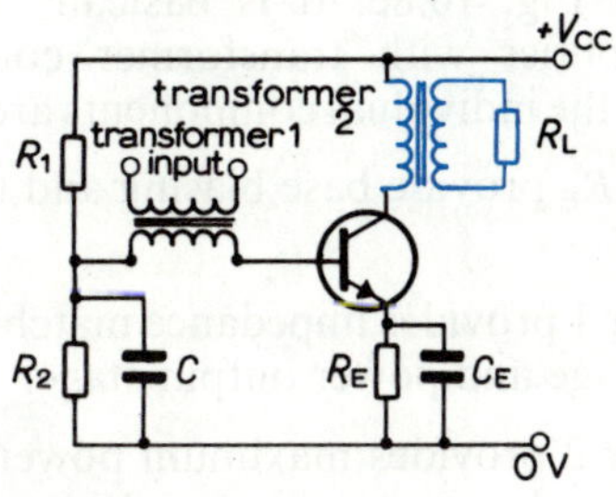

c Transformer coupled power amplifier

Fig. 10.8

circuit, so each point on the curve is determined by the product of I_C and V_{CE} which gives maximum power;

$$P_{(max)} = I_C V_{CE} \qquad (10.11)$$

The blue shading on the output characteristics (Fig. 10.8a) shows the permissible operating region of the transistor to avoid excessive power dissipation and distortion (the signal swing must not enter the non-linear regions of the characteristics).

For maximum efficiency, the load line must touch the maximum power curve. At quiescence (when no signal is present), the transistor is working at its maximum power rating as shown by the product of $I_{CQ}V_{CQ} = P_{(max)}$ in Fig. 10.8a. To help dissipate heat, a heat sink is essential for power transistors.

There are two main disadvantages with this circuit, first, because of the class A biasing, a quiescent collector current flows at all times which is a constant drain on the supply (e.g. reduced battery life). Second, the collector quiescent current flows through the load resistance which is an unwanted dissipation of power (only the signal is required to flow through the load). A direct current in the load may be undesirable. (For example, a loudspeaker may not reproduce sound efficiently because it is saturated by the d.c.)

An improved class A single-ended power amplifier is shown in Fig. 10.8c. It is basically a common-emitter amplifier with transformer coupling. The functions of the individual components are as follows:

R_1, R_2 and R_E provide base biasing and temperature stabilisation.

Transformer 1 provides impedance matching between the driver stage and power output stage.

Transformer 2 provides maximum power transfer to the load, impedance matching and also isolates the load from the quiescent collector current.

Capacitor C_E provides a low impedance path for the signal current to flow in the base-emitter loop.

Class B push-pull amplifier

To supply more power to the load than can be provided by a single-ended amplifier, two transistors are connected in a **push-pull circuit** (Fig. 10.9a).

The class B push-pull amplifier is really two symmetrical circuits. One half amplifies the positive half-cycles of the signal; the other half amplifies the negative half-cycles. The individual outputs are then combined to deliver a composite full-wave to the load.

The windings of the input transformer are so arranged that the bases of the transistors are biased in anti-phase, i.e. during the positive half-cycle T_1 is switched on, T_2 off; during the negative half-cycle T_2 is switched on, T_1 off.

In practice each transistor does not fully conduct until the base-emitter junction barrier voltage is exceeded. The voltage V_B is provided to ensure that each transistor is biased just on. Without this biasing the output waveform would be flattened near zero signal current. This would cause a distortion of the output signal known as **cross-over distortion** (see Fig. 10.9b).

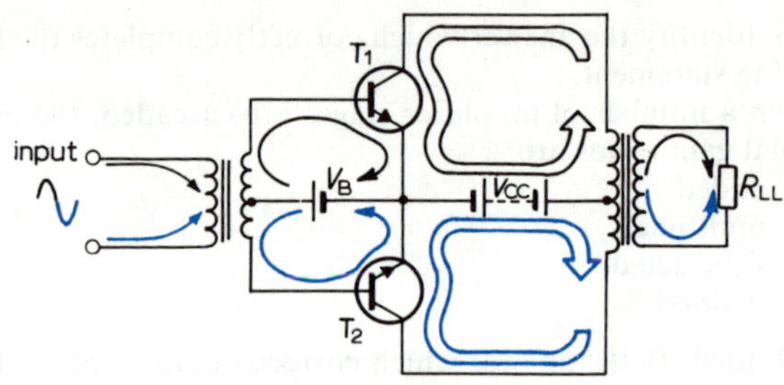

a Class B push-pull circuit

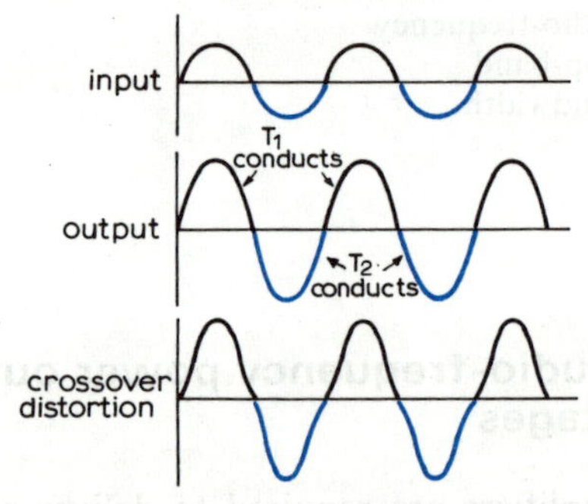

b Waveforms

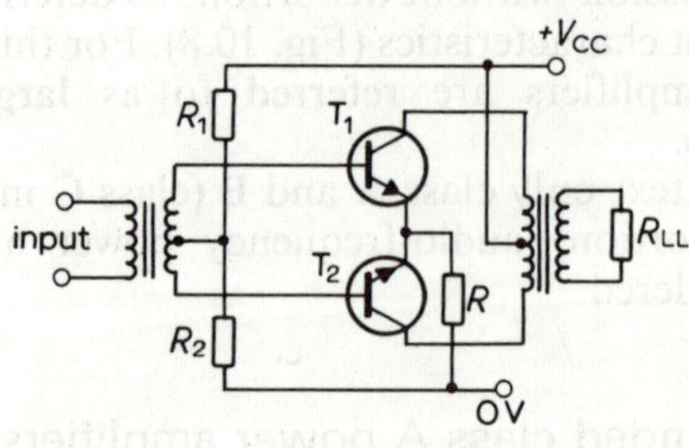

c Practical circuit

Fig. 10.9 Push-pull amplifier

The characteristics of the transistors ideally should be closely matched to provide symmetrical positive and negative half-cycles. A mismatch of current gain would cause one half-cycle to be greater than the other.

A more practical push-pull circuit is shown in Fig. 10.9c. In this circuit the base biasing of the transistors is provided by R_1 and R_2.

The main advantages of the class B push-pull amplifier are economy of power and the reduction of distortion.

Economy of power: The circuit draws little from the supply when no signal is being amplified. Maximum permissible a.c. power output is five times greater than the rated $P_{(max)}$ per transistor.

Reduction of distortion: For example, if mains supply hum is induced in the circuit, each part of the circuit is affected by the same amount. Because of the circuit symmetry, the interference is cancelled before it reaches the load.

Complementary output amplifier

By using transistor **complementary pairs** (one *npn* the other *pnp* with closely matched performance characteristics) push-pull circuits can be designed without the need for transformers (Fig. 10.10a). For this reason they are often referred to as transformerless output stages.

The essential features of a complementary output amplifier are shown in Fig. 10.10a. T_1 is *npn*, T_2 is *pnp*. The emitters are connected together and returned to the mid-point of the supply through the load R_{LL}.

Notice that each transistor is operated in a common-collector configuration (see Section 3.9). Each is biased in class B.

During each positive half-cycle T_1 conducts and T_2 is off (because the base-emitter junction of T_2 is reverse-biased). The potential difference between the collector and emitter of T_1 becomes almost zero, so almost the full positive supply is applied across R_{LL}.

During each negative half-cycle T_2 conducts and T_1 is off. The potential difference across the collector-emitter of T_2 becomes almost zero, so almost the full negative supply is applied across R_{LL}.

The main disadvantage of the circuit shown in Fig. 10.10a is the need for a large input signal to drive current through the load because of the common collector configuration. An improved circuit is shown in Fig. 10.10b where T_2 and T_3 are operated in a common emitter configuration which gives a greater gain.

In this circuit, the input signal voltage is developed

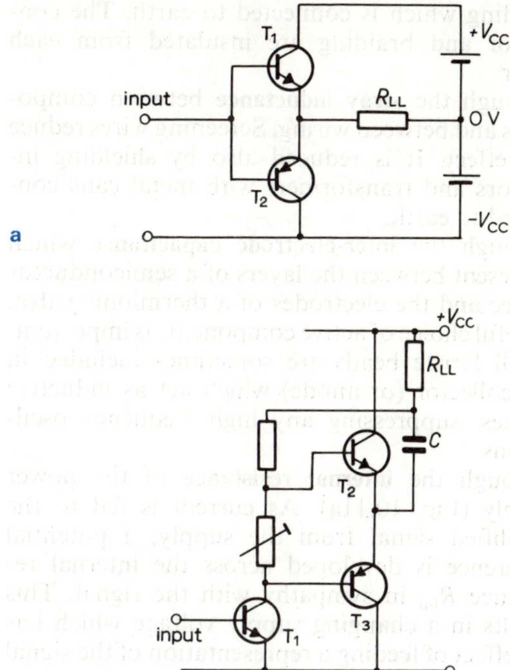

Fig. 10.10 Complementary transistor amplifiers

at the collector of the driver stage T_1, which is applied to the bases of T_2 and T_3. The capacitor C decouples the load resistor R_{LL} from the d.c.

10.12 Parasitic oscillation

Parasitic oscillation is the name given to any unwanted oscillation in an amplifier. It distorts the desired output waveform and causes a waste of power. It is caused by part of the output signal being accidentally re-introduced at the input (this is done purposely for oscillator circuits, see Chapter 7). Power output stages are particularly susceptible to parasitic oscillation because the large signals are easily returned to the input.

Accidental feedback of the signal is possible through a number of paths:

a Through **stray capacitance** between connecting wires. This can be reduced through careful wiring (not having wires with output signals running alongside the input wires). Also by electrically screening suspect wires by using coaxial wire in which the conductor is surrounded by conducting braiding which is connected to earth. The conductor and braiding are insulated from each other.

b Through the **stray inductance** between components and between wiring. Screening wires reduce this effect. It is reduced also by shielding inductors and transformers with metal cans connected to earth.

c Through the **inter-electrode capacitance** which is present between the layers of a semiconductor device and the electrodes of a thermionic valve. Careful choice of active components is important. Small ferrite beads are sometimes included in the collector (or anode) which act as inductive chokes, suppressing any high frequency oscillations.

d Through the **internal resistance** of the power supply (Fig. 10.11a). As current is fed to the amplified signal from the supply, a potential difference is developed across the internal resistance R_{int} in sympathy with the signal. This results in a changing supply voltage which has the effect of feeding a representation of the signal back through the base bias resistors to the inputs of a multi-stage amplifier (the base-bias current depends on the values of R_B and V_{CC}). This may be reduced by the use of appropriately placed **decoupling networks** (Fig. 10.11b). Each series *RC* combination acts as a low impedance filter to earth for the unwanted signals. The resistors R_1 to R_2 ensure that a constant supply is available at each stage of amplification.

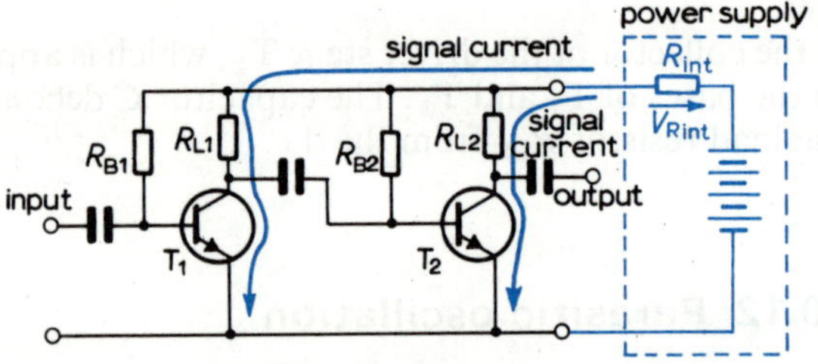

a Signal feedback caused by internal resistance

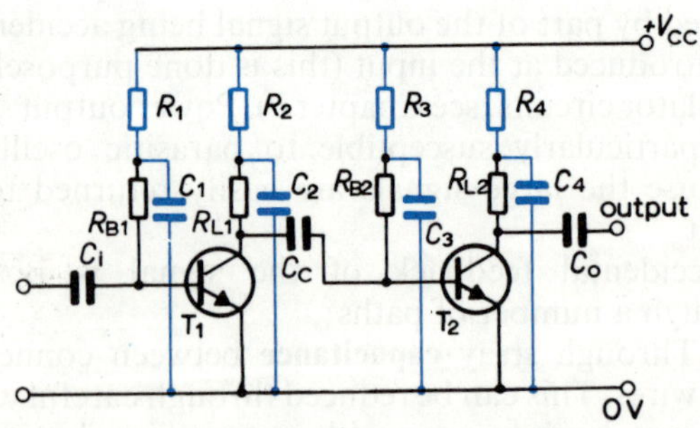

b Decoupling networks

Fig. 10.11

Review

Check your knowledge and understanding. You should:

a Be able to identify single-ended, push-pull, and complementary a.f. large signal amplifiers.

b Understand the function of individual components in the large signal amplifiers of **a**.

c Know the reasons for, and the effects of, parasitic oscillations in electronic amplifiers and how they can be suppressed.

Exercise

10.8 Identify which of the following statements are true and which are false:

a Large signal amplifiers are so called because they use a large proportion of the characteristic curves

b The class A common emitter amplifier can be used as a single-ended large signal amplifier

c Class B push-pull amplifiers are really two amplifiers each amplifying one half cycle of the input waveform

d Complementary amplifiers are used to amplify various parts of the signal frequency then combine those at the output

e A transistor complementary pair eliminates the need for input and output transformers in push-pull circuits

f Parasitic oscillations reduce the gain of large signal amplifiers.

10.9 Identify the answer which correctly completes the following statement. For maximum output efficiency a class A single ended power amplifier requires

a direct coupling to the load

b minimum load impedance

c maximum supply voltage

d transformer coupling to the load.

10.10 Identify the main advantage of the class B push-pull amplifier;

a simplifies the circuit

b economises power

c easier design calculations

d no biasing components required.

10.11 Identify the transistor complementary pair;

a two *npn* transistors

b two *pnp* transistors

c one *npn* and one *pnp* transistor

d one bipolar transistor and one FET.

Self assessment questions

The following are typical of the questions asked to test your achievement of the objectives:

Multiple choice questions

10.12 Identify the class of operation where the full cycle of the input waveform appears at the output.

10.13 If only one half-cycle of the input waveform appears at the output, identify which class the amplifier is operating in

a class A
b class B
c class AB
d class C

10.14
a Identify which of the following list of applications uses class C operation;
b Identify which of the following applications uses class A operation;
i power amplifier
ii a.f. amplifier
iii oscillator
iv tuned r.f. amplifier
v video amplifier

10.15 Identify which of the following statements are true and which are false:
a A tuned amplifier can be used as an a.f. amplifier.
b Selectivity indicates the range of broadcasting stations the receiver can pick up.
c A buffer amplifier is used primarily for inter-stage coupling.
d Class A is used for a.f. voltage amplifiers.
e Class C permits output current flow for less than half the waveform cycle.

10.16 Identify the purpose of impedance matching in amplifier systems;
a to reduce the number of circuit components
b to simplify design calculations
c to bias the amplifying element
d to provide maximum efficiency of amplification

10.17 Identify which type of inter-stage coupling can be used for impedance matching:
a Resistance-capacitive coupling
b Direct coupling
c Transformer coupling
d None

10.18 Assuming the second stage does not affect the first identify the combined gain of a two-stage class A common-emitter amplifier;
a $A_{v1} \times A_{v2}$
b A_{v1}/A_{v2}
c $A_{v1} + A_{v2}$
d $\dfrac{A_{v1} \times A_{v2}}{A_{v1} + A_{v2}}$

10.19 Identify the logarithmic power gain of an amplifier;
a $10 \log_{10} \left(\dfrac{V_o}{V_{in}}\right)$ dB
b $20 \log_{10} \left(\dfrac{V_o}{V_{in}}\right)$ dB
c $20 \log_{10} \left(\dfrac{P_o}{P_{in}}\right)$ dB
d $10 \log_{10} \left(\dfrac{P_o}{P_{in}}\right)^2$ dB

Short answer questions

10.20 With the aid of appropriate diagrams state the conditions for class A, B, AB and C operation of an amplifier and list their individual applications.

10.21 Briefly describe the following types of inter-stage coupling and list their applications:
a resistance-capacitance coupling
b direct coupling
c transformer coupling

10.22 Draw the circuit diagram of a typical radio frequency amplifier. Identify the following parts of the circuit.
a the amplifying element
b the biasing components
c the frequency determining network
d the inter-stage coupling components

10.23 Describe the conditions for maximum efficiency between stages in amplifier systems. State the applications of buffer amplifiers.

10.24 Draw simplified circuit diagrams and list the functions of individual components in the following large-signal amplifiers:
a single-ended
b push-pull
c complementary

10.25 State the reasons for and the effects of parasitic oscillations and methods of suppressing them.

11 Electrical noise

This chapter describes electrical noise, its effects and how these can be minimised. The aims of this chapter are as follows:

- to describe external noise (non-random and random) and to provide and appreciation of its effects
- to describe internal noise (inadequate smoothing, mains hum, thermal or Johnson noise, shot noise, flicker noise, and microphonic noise) and to explain its disadvantages
- to define signal-to-noise ratio.

11.1 Introduction

Ideally any electrical signal being processed, providing some control or conveying information, should be undistorted. In practice all electrical components and devices subject to energy changes generate electrical signals called noise. Electrical noise is an unwanted fluctuation which can interfere with the performance of electronic equipment.

11.2 Effect of electrical noise

Electrical noise is any unwanted fluctuation added to a signal (Fig. 11.1) or any distortion of the signal which tends to mask clear communication. It is generated in all electrical components and is present in any communication or amplifying system.

The effect of noise on an output voltage can be represented diagrammatically as a voltage generator e_n in series with the output (Fig. 11.2). Noise is more apparent when it is mixed with weak signals at an early stage of amplification and then amplified along with the signal. It is noise which limits the useful sensitivity of an amplifier, i.e. the weakest signal which can be amplified and recognised as an independent signal depends on the level of noise present.

A noisy signal may still be recognisable as say speech or music, but it is accompanied by a background hiss, sporadic crackling or the presence of a variable whistle often referred to as **atmospherics**. A television picture may become **snowy**, which is the visual manifestation of electrical noise. Sometimes the speech or music may become distorted as the signal drifts and fades, leaving noise to predominate.

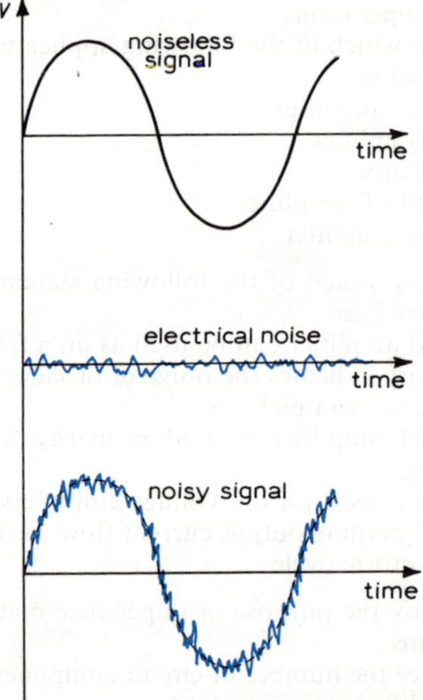

Fig. 11.1 Electrical noise

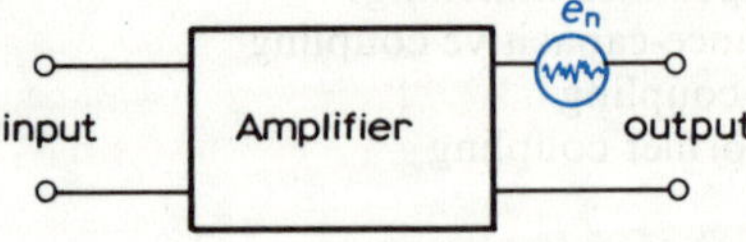

Fig. 11.2 Noise added to amplifier output

Long distance medium wave radio communication is susceptible to this kind of interference.

11.3 External noise

Electrical noise can be considered as **external noise**, generated outside the receiver or amplifier, or **internal noise**, the source being within the receiver or amplifier.

External noise can be non-random (predictable noise), or random (non-predictable noise).

Non-random noise: This follows an identifiable regularity and is normally man-made, caused by the accidental electro-magnetic radiation of machines such as electric motors and generators.

Measures can be taken to reduce this sort of radiated noise by screening the source or suppressing it with a suitable capacitor. By law, car ignition circuits have to be suppressed. You may have noticed the sharp crackle on a radio or television when a poorly suppressed vehicle passes close by the aerial. The radiation from such sources is usually spread over a very wide frequency range, even though its amplitude variation may be low frequency.

If the noise has a constant low frequency which is substantially lower than the signal frequency, much of the noise energy can be short-circuited to earth by connecting a capacitor across the input of the receiver. The value of the capacitance is chosen such that its reactance is very low at the noise frequency. All other frequencies (i.e. the signal frequencies) are blocked from this path to earth and have to pass through the receiver.

Random noise: This source of external random noise is often found in heavenly bodies, particularly those of a very high temperature such as the stars. The sun is our strongest source because of its closeness. Violent sunspot activity (which follows a cycle of eleven years) causes a great deal of disturbance in the Earth's atmosphere. Long range communication has been known to be blacked out by these disturbances for many hours.

An American researcher, Karl Guth Janskey, first recognised the extra-terrestrial radio emissions in 1932 whilst investigating the cause of interference on trans-atlantic telephone cables. Before this, no-one had realised that space was full of electro-magnetic radiation carrying a vast resevoir of information about the structure and origin of the universe. Radio astronomy as we know it today developed from an investigation into these effects.

A recieved radio signal can drift or fade. This kind of interference mainly affects long distance communication which depends on reflection from the Earth's ionisation layer in the upper atmosphere. When the reflective ability of the ionisation layer fluctuates (this is caused by clouds of ionised particles drifting through the upper atmosphere), the range of the reflected signal varies.

During fading, no additional noise is present, but as the signal becomes weaker, the **signal to noise ratio** (see Section 11.5) is degraded.

Although external noise can be reduced, methods of eliminating it are limited because this type of noise is often a composite part of the signal arriving at the receiver. Carefully designed receiving aerials with directional qualities reduce noise from unwanted directions. This is the purpose of the reflecter and director elements attached to television aerials.

High quality FM transmissions, such as for stereo broadcasts, use a system of **pre-emphasis** and **de-emphasis** of the audio signal. Pre-emphasis is the increase of the audio above 10 kHz. Because most of the external noise is added to the high frequencies, de-emphasis at the receiver reduces the noise as the audio frequencies are restored to their normal level.

Digital transmission techniques (e.g. pulse code modulation see Chapter 8) are now often used to reduce noise interference.

Review

Check your knowledge and understanding. You should:

a Be able to appreciate the meaning of electrical noise.

b Know the sources of external noise.

c Know the precautions which can be taken to minimise the effects of external noise.

11.4 Internal noise

Any amplifier inherently adds noise to the signal. Even when no signal is present at the input, a noise output can be detected. This is known as internal or amplifier noise.

Internal noise is generated by one or more of the following;

a inadequate smoothing of the power supply

b mains hum pick up

c thermally generated noise (Johnson or white noise)

d irregularity in the motion of charge carriers (shot noise)

e surface effects in semiconductor and cathode emitting materials (flicker noise)
f mechanical vibration of the electrodes in thermionic valves (microphonic noise)

Each of these is discussed in greater detail in the following sections.

Smoothing

If a power supply smoothing circuit is inadequate, the amplifier supply voltage fluctuates undesirably.

From Eqn. 4.1 of Section 4.2, the base quiescent current of a common-emitter amplifier depends on V_{CC}, that is,

$$I_{BQ} = \frac{V_{CC} - V_{BE}}{R_B}$$

If V_{CC} varies, so does I_{BQ} which is amplified and the fluctuation is superimposed on the wanted signal. Careful design of the power supply can reduce this source of noise to a minimum (see Chapter 13. Stabilised power supplies).

Mains hum

In poorly designed amplifiers, mutual inductive coupling may link the a.c. power supply to the amplifier circuits. The induced e.m.f. causes noise in the form of **mains hum** which has a frequency of 50 Hz. This can be reduced to a minimum by careful design to eliminate the inductive coupling, by for example, shielding the power supply with a screen of metal.

Numerous common connections or earth connections can cause noise when the earth runs have resistance between the connecting points. This means that a p.d. is developed between the earth points. These can be eliminated by using low resistance paths to a common earth point.

Thermal or Johnson noise

Random fluctuations of electrons in a conductor or any electrical component are caused by the absorption of heat energy. The fluctuations are very small, but they are sufficient to generate small noise potentials. The resulting noise is called **thermal** or **Johnson noise**.

There are multitudes of electrons in a conductor or component contributing to the total noise level. Because the noise is generated evenly over all frequencies, it is also known as **white noise**, (which is analogous to white light containing all the visible frequencies).

The r.m.s. value of the thermal noise voltage V_n in a resistor R over a frequency range $f_2 - f_1$ is given by the relationship

$$V_n{}^2 = 4kTRB \qquad (11.1)$$

where k is Boltzmann's constant (a constant of proportionality which equals 1.38×10^{-23} J °K^{-1}),
T is absolute temperature of the resistor, °K,
R is the resistance in ohms,
and B is the equivalent noise bandwidth, $f_2 - f_1$ Hz.

This means that a 100 kΩ resistor at room temperature (300° K) over a bandwidth of 10 kHz generates a noise voltage V_n of 4.0 μV.

Because the noise voltage is directly proportional to the bandwidth, it is wise to design low noise, high gain amplifiers with the minimum possible bandwidth yet preserving all the harmonic content of the signal.

For applications such as the reception of satellite communication signals or the monitoring of weak radio signals from deep space, the amplifiers are often cooled to a very low temperature, with liquid nitrogen (−196°C), to reduce the thermal noise.

Shot noise

Shot noise is caused by the irregularities of motion of individual electrons in a current. There is a tendency to consider current as being a consistently smooth flow when in fact, it is the transition of millions of small charged particles. Electron flow has inherent irregularities. The passage of electrons through amplifying devices such as the bipolar transistor, FET and thermionic valve, is erratic. These irregularities of a current (shot noise) are impressed on the signal being amplified.

Flicker noise

Flicker noise is the common name given to two distinct phenomena, but each manifesting noise in a similar manner. One is the spontaneous emission of charged particles from an oxide-coated cathode giving rise to noise. The other is thought to be the effect of electron-hole pair generation and recombination on the surface of a semiconductor.

At low frequencies (below about 1 kHz), flicker noise varies approximately as the reciprocal of frequency and is often referred to as $1/f$ noise.

Microphonic noise

The amplifying performance of thermionic valves is greatly dependent upon the distances between the electrodes. If these distances fluctuate, even slightly, because of mechanical or acoustic vibration, a noise signal is generated, generally called **microphonic noise**.

Microphonic noise is minimised by mounting valves on rubber mountings to absorb vibration.

Valves with many electrodes are more susceptible to microphonic noise. Therefore triodes are often preferred to the pentode (five electrodes) for low noise applications.

Review

Check your knowledge and understanding. You should:

a Know the sources of internal noise in electronic devices.

b Appreciate some of the ways of minimising them.

Exercise

11.1 Identify which of the following statements are true and which are false:

a Electrical noise can be defined as any unwanted voltage added to a signal.

b All external noise can be reduced almost completely in an amplifier.

c Microphonic noise occurs only in valves.

d White noise or Johnson noise affects only the visible part of the electromagnetic spectrum.

e Electrical noise causes fading in long distance communication.

11.2 Calculate the value of white noise voltage developed in a 1 MΩ resistor over a bandwidth of 20 kHz at

a Room temperature (300° K)

b absolute zero (0° K)

c the temperature of liquid nitrogen (77° K)

d at boiling point of water (373° K)

11.5 Signal to noise ratio

The ratio of signal power to noise power is a useful measurement of noise content. The individual magnitudes of the signal and noise are not as important as the relative magnitudes. So long as the signal is sufficiently larger than the noise, it can be amplified and identified.

$$\text{Signal/noise power ratio} = \frac{S}{N} \quad \text{(no units)} \tag{11.2}$$

or, Signal/noise power ratio (in decibels)

$$= 10 \log_{10} \frac{S}{N} \text{ dB} \tag{11.3}$$

where S is the signal power and N is the noise power present in the signal.

Because both signals are assumed to be present in the same resistance R, the r.m.s. voltages can be related to power $\left(P = \frac{V_{(\text{r·m·s·})}^{2}}{R}\right)$

$$S = \frac{V_s^2}{R}$$

$$N = \frac{V_n^2}{R}$$

$$\text{Signal/noise power ratio} = \frac{V_s^2}{R} \times \frac{R}{V_n^2}$$

$$= \left(\frac{V_s}{V_n}\right)^2 \tag{11.4}$$

Signal/noise power ratio (in decibels)

$$= 10 \log_{10} \left(\frac{V_s}{V_n}\right)^2$$

$$= 20 \log_{10} \left(\frac{V_s}{V_n}\right) \text{ dB} \tag{11.5}$$

Review

Check your knowledge and understanding. You should:

a Be able to define signal to noise ratio in an amplifier or receiver.

b Be able to calculate the signal to noise ratio, in dB, given signal and noise powers for a particular device.

Exercise

11.3 Calculate the signal to noise power ratio at the output of an amplifier when the signal power is 5 W and the noise power is 3 mW.

11.4 Calculate the signal to noise power ratio at the output of an amplifier when the r.m.s. signal voltage is 8 V and the r.m.s. noise voltage is 10 mV.

11.5 If the output of an amplifier to one loudspeaker is 10 W and the signal to noise power ratio is 50 dB calculate

a the noise power

b the noise peak voltage when the peak signal voltage is 8 V.

Self assessment questions

The following are typical of questions you may be asked in an assessment test.

Multiple choice questions

11.6 Identify which of the following are external sources of noise and which are internal;

a mains hum
b rotating electrical machinery
c flicker noise
d microphonic noise
e sun and stars
f motor car ignition
g shot noise
h Johnson or white noise

11.7 Identify which type of internal noise each of the following describes;

a noise generated by electron-hole pair generation at surface of a semi-conductor
b noise generated by random fluctuation of electrons in a conductor
c noise generated by the mechanical vibration of the electrodes of a thermionic valve
d noise generated by the spontaneous emisson of a charge particle from an oxide-coated cathode
e noise generated by the uneven distribution of electrons in current

11.8 Identify which of the following is an expression for signal to noise power ratio;

a $20 \log_{10}\left(\frac{S}{N}\right)$ dB

b $20 \log_{10}\left(\frac{V_s}{V_n}\right)$ dB

c $10 \log_{10}\left(\frac{S}{N}\right)^2$ dB

d $20 \log_{10}\left(\frac{V_s}{V_n}\right)^2$ dB

Short answer questions

11.9 List sources of external and internal noise. Briefly describe the sources of each.

11.10 State precautions taken to minimise external noise.

11.11 Calculate the signal to noise power ratio at the output of an amplifier when the signal power is 4 W and the noise power is 3 mW.

12 Feedback amplifiers

This chapter introduces the general principles of feedback amplifiers. The aims of this chapter are as follows:

- to provide an appreciation of the concept of feedback in electronics
- to provide an understanding of the effect of feedback in an amplifier (gain, gain stability, bandwidth, distortion, noise, and input and output resistances)
- to describe practical feedback circuits for improving amplifier performance.

12.1 Introduction

Ideally the output signal of an amplifier should be a faithful reproduction of the input, with a larger amplitude. The amplifiers considered in previous chapters provide outputs which are not faithful reproductions of the input. Degradation is a result of electrical noise (as considered in the previous chapter), and the non-linearity of the amplifier circuit and components (they vary in performance for different conditions).

Improvement in the output signal of an amplifier can be accomplished by means of compensating circuits which generally use part of the output signal in comparison with the input signal to change the response of the amplifier. Such circuits provide a link between the output and input and consequently are called **feedback circuits**. Amplifiers employing such compensation are called **feedback amplifiers**.

12.2 Concept of feedback

Information from the output signal of an amplifier can be fed back to modify the input signal. Such modification changes the characteristics of the amplifier. The nature of the feedback can improve aspects of the amplifier performance.

Fig. 12.1a shows a block diagram of an amplifier with an added feedback network (circuit).

The amplifier, of gain A_v, on its own is called an **open-loop amplifier**. The total system with a loop connecting the output and input is called a **closed-loop amplifier**. The elements of the feedback amplifier in Fig. 12.1a are amplifier A_v, feedback network β, sampling junction and mixing junction.

The open-loop amplifier provides an **open-loop gain**, A_v.

Feedback networks take many forms, some of which are described in Section 12.6. They provide the path through which a signal derived from the output is fed back to the input. The feedback network may, or may not invert the signal from the sampling junction, depending upon the design requirements of the system. (In practice the input impedance of the feedback

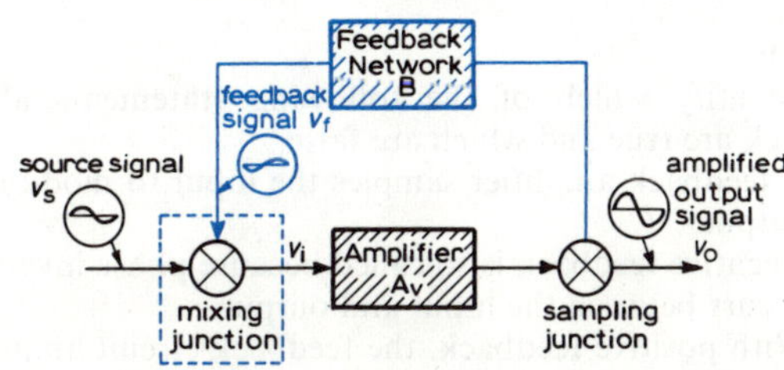

a Block diagram of feedback amplifier

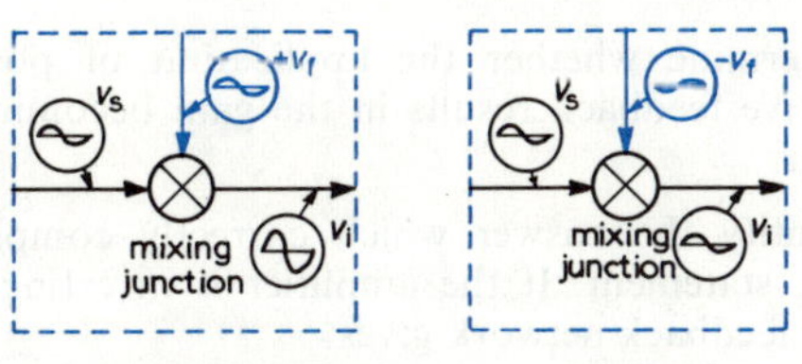

b Positive feedback   c Negative feedback

Fig. 12.1 The principles of feedback

network needs to be much higher than the amplifier load impedance to ensure that the additional loading of the amplifier is minimal.)

The **sampling junction** is that part of the output circuit from which the feedback information is derived.

The **mixing junction** combines the feedback signal with the input signal.

The feedback signal can be applied so that it either adds to the input signal (Fig. 12.1b), which normally increases the overall gain, or subtracts from it (Fig. 12.1c), which reduces the overall gain.

If the overall gain (v_o/v_s) is increased, the feedback is said to be regenerative or **positive feedback**. If the overall gain is decreased the feedback is said to be degenerative or **negative feedback**.

Review

Check your knowledge and understanding. You should:

a Know how to draw the block diagram of a feedback amplifier system.
b Be able to define the elements of a feedback amplifier system.
c Know what is meant by positive and negative feedback in amplifiers.

Exercise

12.1 Identify which of the following statements about feedback are true and which are false.

a A feedback amplifier samples the input to modify the output.
b Negative feedback is so called because phase inversion occurs between the input and output.
c With positive feedback, the feedback circuit amplifies the signal.
d Negative feedback descreases the overall gain of an amplifier.
e A feedback amplifier samples the output to modify the input.

12.2 Determine whether the application of positive or regenerative feedback results in the gain becoming larger or smaller.

12.3 Identify the answer which correctly completes the following statement. If the amplifier is inverting, a non-inverting feedback network gives

a positive feedback
b negative feedback
c could be positive or negative feedback
d no feedback at all

12.3 Stage gain of a feedback amplifier

Modification of the gain of an amplifier with the introduction of a feedback network can be seen from an expression for the **closed-loop gain** (A_{vf}) of the amplifier in terms of A_v (the **open-loop gain**), and β (the **feedback factor**). Closed-loop gain is obtained from the ratio v_o/v_s, that is

$$A_{vf} = \frac{v_o}{v_s} \tag{12.1}$$

where v_o is the output signal voltage and v_s is the input signal voltage. The feedback factor β determines the fraction of the output voltage fed back to the input. Hence

$$v_f = \beta v_o \tag{12.2}$$

where v_f is the feedback voltage. β has a value between zero and unity. When $\beta = 0$, no feedback is applied. When $\beta = 1$, 100 per cent feedback is applied (all the output voltage).

When feedback is applied, the input voltage v_i is the sum of v_s and v_f,

$$v_i = v_s + v_f \tag{12.3}$$

The input voltage of the amplifier v_i is amplified by the open-loop gain A_v

$$v_o = A_v v_i$$

Substituting Eqn. 12.3, $v_o = A_v(v_s + v_f)$ (12.4)

Substituting Eqn. 12.2 for v_f

$$\begin{aligned} v_o &= A_v(v_s + \beta v_o) \\ &= A_v v_s + A_v \beta v_o \\ v_o - A_v\beta v_o &= A_v v_s \qquad (12.5) \\ v_o(1 - A_v\beta) &= A_v v_s \end{aligned}$$

$$\therefore A_{vf} = \frac{v_o}{v_s} = \frac{A_v}{(1 - A_v\beta)} \tag{12.6}$$

Eqn. 12.6 is the stage gain relationship for an amplifier with feedback. (It is discussed in detail when considering negative feedback (Section 12.4) and positive feedback (Section 12.5).

If A_v is so large that $A_v\beta \gg 1$, then the 1 can be ignored and Eqn. 12.6 becomes

$$\begin{aligned} A_{vf} &\approx -\frac{A_v}{\beta A_v} \\ &\approx -\frac{1}{\beta} \qquad (12.7) \end{aligned}$$

Review
Check your knowledge and understanding. You should:

Know how to derive the general expression for stage gain of a feedback amplifier.

12.4 Negative feedback

Although negative feedback causes a reduction of gain, it is widely used for its many advantages; improved stability of gain, increase in bandwidth, reduction of noise and distortion, and change in input and output impedances.

Reduction of gain

From Eqn. 12.6 we can deduce that when $\beta = 0$, no feedback is applied and the gain reverts to A_v.

For negative feedback, $A_v\beta$ becomes negative because v_f is negative as shown in Eqn. 12.4. Therefore Eqn. 12.6 becomes

$$A_{vf} = \frac{A_v}{(1 + A_v\beta)} \qquad (12.8)$$

When β lies between the values of zero and unity ($0 \leqslant \beta \leqslant 1$), the denominator term $(1 + A_v\beta)$ becomes greater than unity. The closed-loop gain A_{vf} must then be lower than the open-loop gain A_v.

Example 12.1
A common-emitter amplifier has a mid-frequency open-loop voltage gain of $A_v = -10^4$ (note that the negative sign denotes phase inversion). If 2% negative feedback is applied, determine the gain with feedback A_{vf}.
Using Eqn. 12.6

$$A_{vf} = \frac{-10^4}{1 - [(-10^4) \times 0.02]} = -\frac{10^4}{1 + 200}$$
$$= -49.75$$

Note; negative feedback is achieved by an inverting amplifier and not by phase reversal due to the 2% negative feedback.

Stability of gain

Any fluctuation in the open-loop gain is reduced by negative feedback by the same proportion as the reduction in gain, that is, if A_v varies by 10% and A_{vf} is one-fifth of A_v then A_{vf} varies by $\frac{10}{5}\% = 2\%$.

Another way of writing this is

$$\text{variation of } A_{vf} = \frac{\text{variation of } A_v}{(1 + A_v\beta)} \qquad (12.9)$$

This reduction is *not* simply a reduction in magnitude but a reduction of the percentage variation as shown by plotting the percentage variation against gain in Fig. 12.2a Although the open-loop gain may vary considerably, the closed-loop gain is affected to a lesser degree. From this it can be concluded that an amplifier becomes more stable with the application of negative feedback.

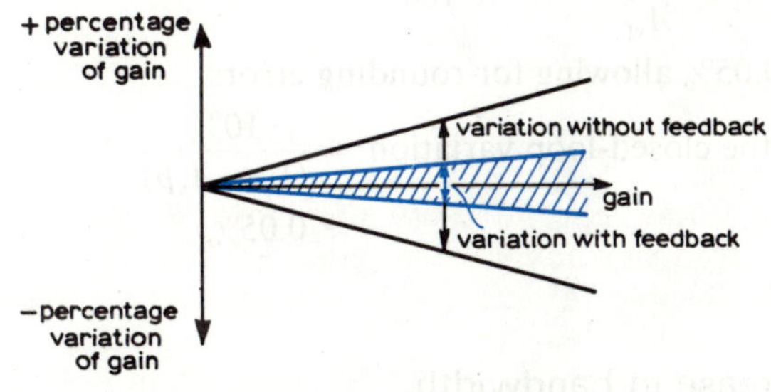

a Reduction of gain variation

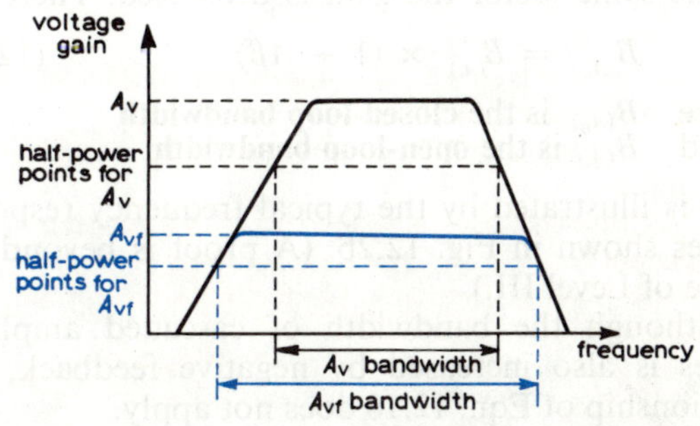

b Increase in bandwidth

Fig. 12.2

Example 12.2
If the open-loop gain of Example 12.1 varied by ±10%, determine the resultant percentage variation when 2% negative feedback is applied.

The open-loop gain $A_v = -10^4 \pm 10\%$

$$\text{Upper } A_v = -10^4 - 10^3 = -11 \times 10^3$$

$$\text{Upper } A_{vf} = \frac{-11 \times 10^3}{(1 + 11 \times 10^3 \times 0.02)} = -49.77$$

$$\text{Lower } A_v = -10^4 + 10^3 = -9 \times 10^3$$

$$\text{Lower } A_{vf} = \frac{-9 \times 10^3}{(1 + 9 \times 10^3 \times 0.02)} = -49.72$$

$$A_{vf} = -49.75 \text{ (from Example 12.1).}$$

The closed loop percentage variation

$$= \frac{\text{upper } A_{vf} - A_{vf}}{A_{vf}} \times 100$$

$\approx 0.05\%$ allowing for rounding errors.

$$\text{or, the closed-loop variation} = \frac{10\%}{(1 + A_v\beta)} \approx 0.05\%$$

Increase in bandwidth

The bandwidth (B) of a single-stage amplifier is increased with the application of negative feedback by the same factor the gain is decreased. Therefore

$$B_{(A_{vf})} = B_{(A_v)} \times (1 + A\beta) \qquad (12.10)$$

where $B_{(A_{vf})}$ is the **closed-loop bandwidth**
and $B_{(A_v)}$ is the **open-loop bandwidth.**

This is illustrated by the typical frequency response curves shown in Fig. 12.2b. (A proof is beyond the scope of Level III.)

Although the bandwidth of cascaded amplifier stages is also increased by negative feedback, the relationship of Eqn. 12.10 does not apply.

Reduction of noise and distortion

With reference to Fig. 12.3, consider how internal noise is reduced by the application of negative feedback. For simplification, initially assume no input signal is present.

Noise that is generated within the amplifier appears amplified at the output (Fig. 12.3a). The feedback network returns a reduced representation of the noisy

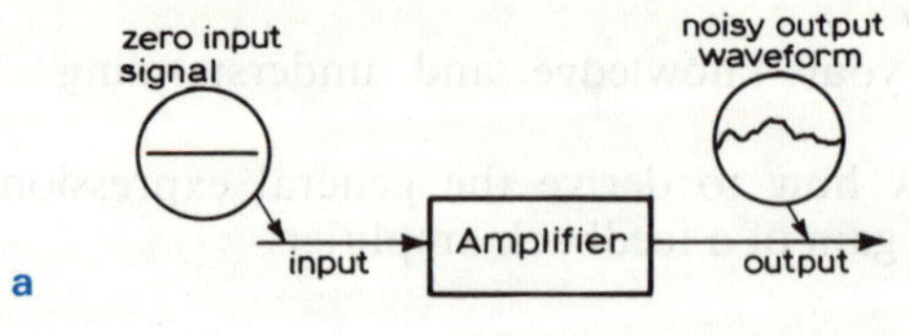

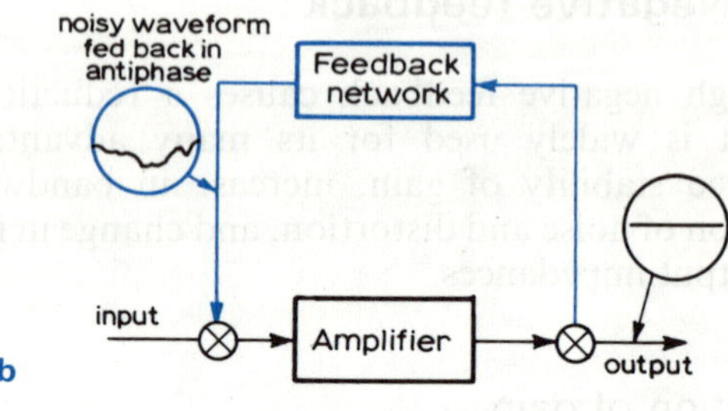

Fig. 12.3 Reduction of internal noise

output in antiphase. This is passed through the amplifier, tending to cancel the original noise (Fig. 12.3b).

Noise generated in the early stages of an amplifier is amplified by subsequent stages so it is very important to reduce it. (Note that external noise is not reduced by negative feedback as this is a composite part of the input signal.)

Noise superimposed on the wanted input signal is reduced in a similar manner; any deviation of the output waveform from the input waveform is reduced. This includes any distortion caused by the non-linear operation of active devices.

In practice, the cancellation of internal noise is not complete, but is reduced considerably. With negative feedback, the signal/noise ratio is improved by the same factor by which the gain is reduced.

$$\frac{S_{(fb)}}{N} = \frac{S}{N} \times (1 + A_v\beta) \qquad (12.11)$$

Input and output impedance

The effect of negative feedback on the input impedance of an amplifier depends on how the feedback is applied. If the feedback signal is introduced to the input in series (Fig. 12.4a and c), the input impedance is increased. This is because the input voltage must be increased to maintain the same input current. Conversely, if the feedback signal is introduced to the input in shunt or parallel (Fig. 12.4b and d), the input impedance is reduced. This is because the impedance of the feedback network is parallel with the amplifier impedance.

Generally:

for **series feedback** $Z_{i(fb)} = Z_i(1 + A_v\beta)$ (12.12)

and for **shunt feedback** $Z_{i(fb)} = \dfrac{Z_i}{(1 + A_v\beta)}$ (12.13)

where Z_i is the input impedance without feedback, and $Z_{i(fb)}$ is the input impedance with feedback.

The effect of negative feedback on output impedance depends on whether the feedback signal is derived from the output voltage or the output current.

Generally:

for output **current-derived feedback** $Z_{o(fb)}$

$$= Z_o(1 + A_v\beta) \quad (12.14)$$

for output **voltage-derived feedback** $Z_{o(fb)}$

$$= \frac{Z_o}{(1 + A_v\beta)} \quad (12.15)$$

where Z_o is the output impedance without feedback, and $Z_{o(fb)}$ is the output impedance with feedback.

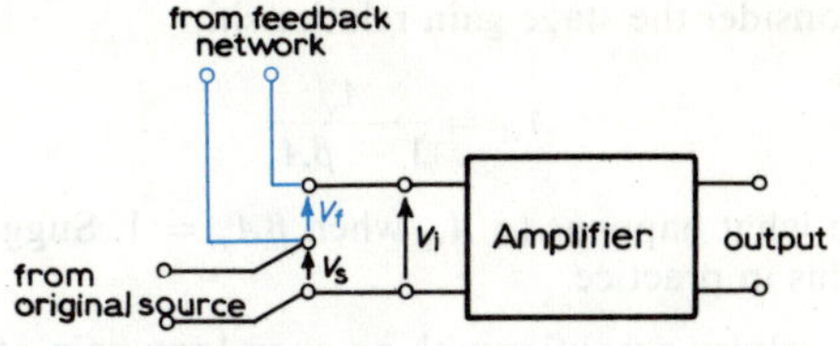

a Voltage fed back in series

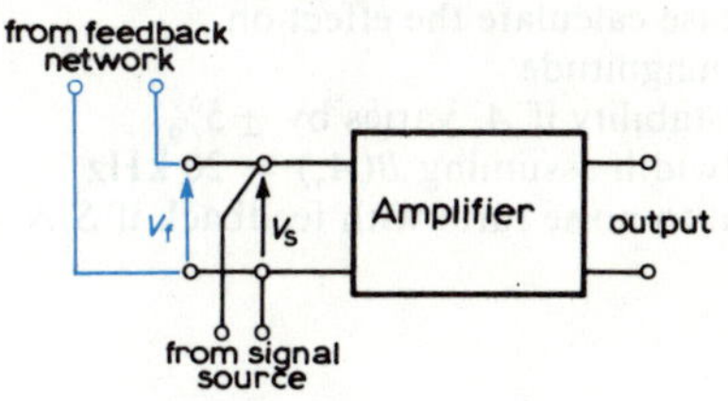

b Voltage fed back in shunt

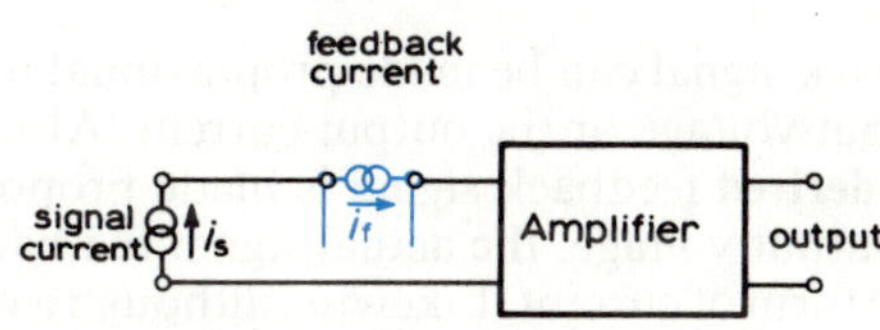

c Current fed back in series

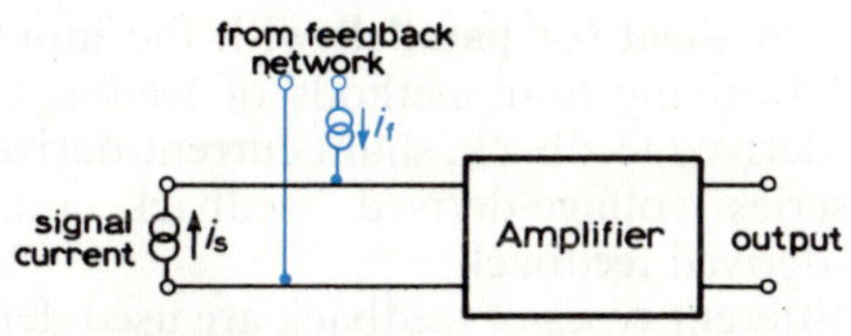

d Current fed back in shunt

Fig. 12.4 Four types of feedback signal

12.5 Positive feedback

In the derivation of Eqn. 12.6, it has been assumed that the feedback is regenerative or positive.

Close examination of Eqn. 12.6 reveals that if $\beta = 0$, the gain reverts to the open-loop gain A_v. For some values of the feedback factor β, the denominator term is reduced below unity and A_{vf} becomes greater than A_v, i.e. the gain is increased by positive feedback.

This may initially sound desirable, but the increase in gain is accompanied by certain instabilities which distort the output signal or even cause oscillation. Hence oscillator circuits generally are positive feedback amplifiers (see Chapter 7).

Positive feedback has the opposite effects to those of negative feedback.

Review

Check your knowledge and understanding. You should be able to understand the effects of applying negative feedback to an amplifier in relation to gain, gain stability, bandwidth, distortion, noise, and input and output impedance.

Exercise

12.4 Define the difference between the open-loop gain A_v and the closed-loop gain A_{vf} of an amplifier.

12.5 Identify which of the following statements is true when comparing A_v with A_{vf}:

- **a** A_{vf} is always greater than A_v
- **b** A_{vf} is always less than A_v
- **c** A_{vf} is always equal to A_v
- **d** The relative sizes of A_v and A_{vf} depend on whether positive or negative feedback is applied.

12.6 Consider the stage-gain relationship

$$A_{vf} = \frac{A_v}{1 - \beta A_v}$$

Explain what happens to A_{vf} when $\beta A_v = 1$. Suggest what limits this in practice.

12.7 A voltage amplifier with an open loop gain of 10^4 has applied to it in turn
- **a** 1% negative feedback
- **b** 3% negative feedback

For each case calculate the effect on
- **i** gain magnitude
- **ii** gain stability if A_v varies by $\pm 5\%$
- **iii** bandwidth assuming $B(A_v) = 20$ kHz
- **iv** signal to noise ratio with feedback if $S/N = 30$ dB.

12.6 Feedback circuits

A feedback signal can be made proportional to either the output voltage or the output current. Although a **voltage-derived** feedback signal is made proportional to the output voltage, the actual signal fed back may be in the form of current. Likewise, although a **current-derived** feedback signal is made proportional to the output current, the actual signal fed back to the input may be a voltage.

The feedback signals can be fed back either in **series** or in **shunt** (or **parallel**) with the input signal (Fig. 12.4) giving four methods of feedback; series current-derived feedback, shunt current-derived feedback, series voltage-derived feedback and shunt voltage-derived feedback.

The different types of feedback are used depending on requirements of performance to give the required gain, stability and input and output impedances. **Hybrid** or **compound feedback** circuits using more than one of the above methods are sometimes employed in special applications.

Series current-derived feedback

A block diagram of the **series current-derived feedback** amplifier is shown in Fig. 12.5a. The feedback (a voltage fed back in series) is derived from the output current flowing through the feedback resistor R_f. The feedback voltage (v_f) opposes the signal voltage (v_s), hence negative feedback. An interchange of connections to R_f would make the feedback positive.

Fig. 12.5b shows how series current-derived feedback can be implemented in a practical circuit. You may recognise the common-emitter amplifier with base potential-divider biasing (see Section 4.7.) The current flowing through the feedback resistor R_f is i_E which is approximately the same as the output current i_C. The voltage v_f reduces the base-emitter voltage v_{BE}. Note that the base-emitter voltage and v_f are in series.

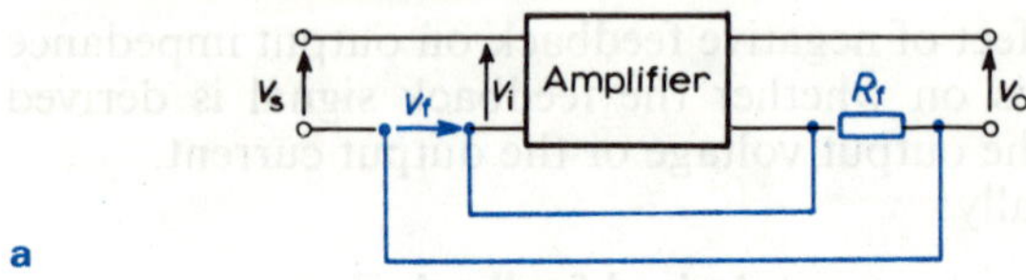

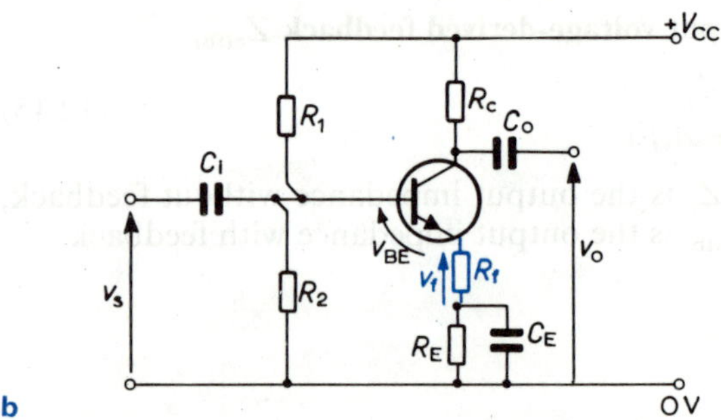

Fig. 12.5 Series current-derived feedback

Shunt current-derived feedback

A block diagram of the **shunt current-derived** feedback amplifier is shown in Fig. 12.6a. The feedback signal (a voltage fed back in shunt) is derived from the output current.

A practical form of shunt current-derived feedback is shown in Fig. 12.6b. The feedback signal is a voltage developed across feedback resistor R_f, which is then fed back in shunt with the input. The feedback signal is proportional to i_{E2} which is almost the same magnitude as i_{C2}.

A signal taken from the emitter of a common-emitter amplifier as in Fig. 12.6b, is in phase with the input of the stage. This is so because the p.d. across R_E varies in phase with i_E and i_C which follows the input signal. Because the input to T_2 already has been inverted by the first stage, the feedback signal is out of phase with v_{s1} giving negative feedback.

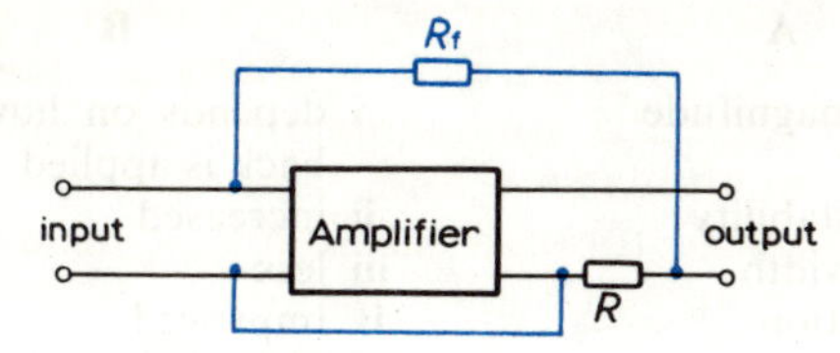

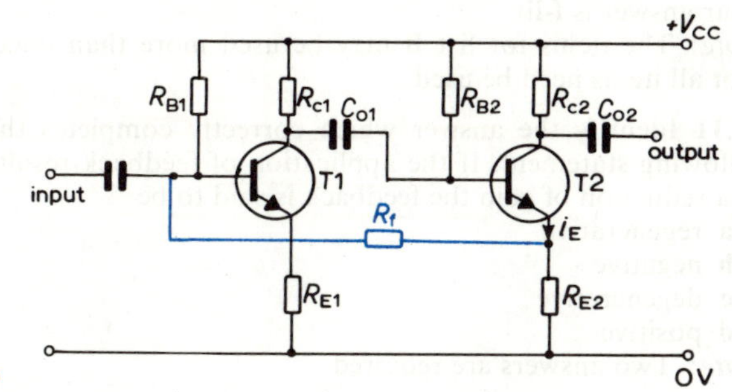

Fig. 12.6 Shunt current-derived feedback

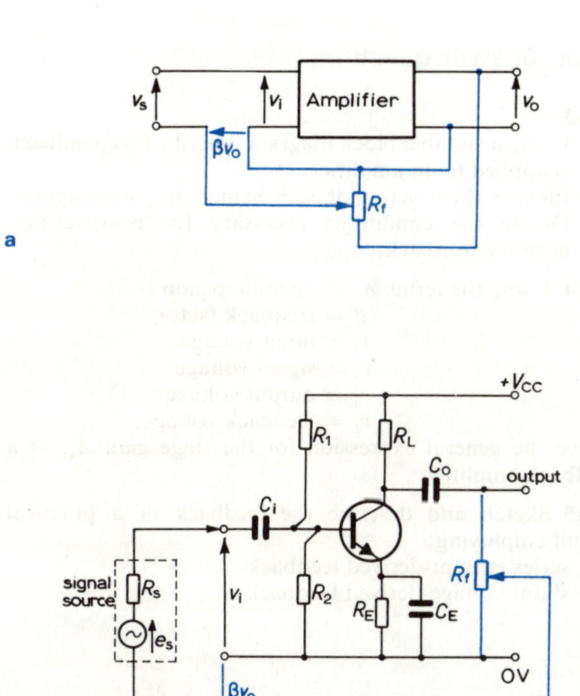

Fig. 12.7 Series voltage-derived feedback

Series voltage-derived feedback

A block diagram of the **series voltage-derived feedback** amplifier is shown in Fig. 12.7a. A practical method of implementing this is shown in Fig. 12.7b. The amount of feedback signal βv_o is determined by the position of the slider of potentiometer R_f. The signal voltage v_s and βv_o are in series and because the common-emitter amplifier is inverting, the feedback is negative.

Shunt voltage-derived feedback

A block diagram of the **shunt voltage-derived feedback** amplifier is shown in Fig. 12.8a. Resistor R_f provides the path between input and output. The feedback signal is derived from the voltage at the output, although the signal itself is actually a current proportional to v_o. The feedback signal is in shunt with the input. A simple practical circuit is shown in Fig. 12.8b.

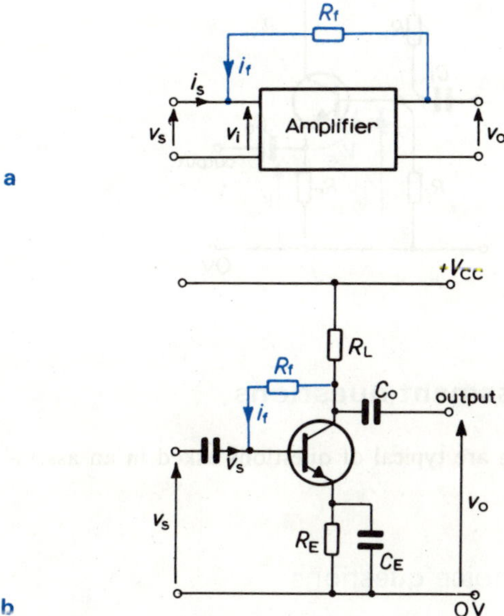

Fig. 12.8 Shunt voltage-derived feedback

Review

Check your knowledge and understanding. You should be able to draw block diagrams and examples of practical circuits for series current, shunt current, series voltage and shunt voltage-derived feedback.

Exercise

12.8 Identify the answer which correctly completes the following statement. If a current proportional to the output voltage is back in parallel with the input, the feedback is said to be;

a series current-derived feedback
b shunt current-derived feedback
c series voltage-derived feedback
d shunt voltage-derived feedback

12.9 Consider the action of the emitter follower circuit of Fig. 12.9.
The voltage across the transister V_{BE} controls the amplification of the circuit. Because V_{BE} is the difference between V_o and V_{in} 100% feedback is achieved.
Now answer the following questions:

a Is the feedback signal proportional to the output current or voltage?
b Is the feedback signal in series or shunt with the input?
c The type of feedback is;
  i series current-derived feedback
  ii shunt current-derived feedback
  iii series voltage-derived feedback
  iv shunt voltage-derived feedback

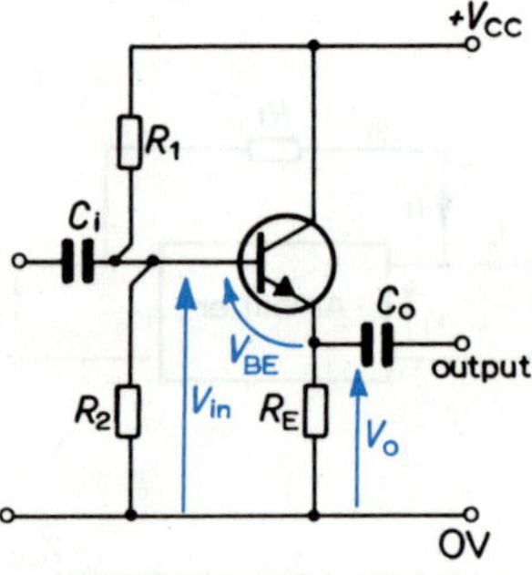

Fig. 12.9

Self assessment questions

The following are typical of questions asked in an assessment test:

Mutliple choice questions

12.10 Match the letters and the numbers of the two lists below to describe the effects of negative feedback on amplifier performance.

A	B
a gain magnitude	**i** depends on how feedback is applied
b gain stability	**ii** increased
c bandwidth	**iii** less
d distortion	**iv** improved
e noise	**v** degraded
f input and output impedance	**vi** smaller

(e.g. if you think the input and output impedance is increased your answer is **f-ii**).
Note: The items for list B may be used more than once. Not all items need be used.

12.11 Identify the answer which correctly completes the following statement. If the application of feedback results in a reduction of gain the feedback is said to be:

a regenerative
b negative
c degenerative
d positive

Note: Two answers are required

12.12 Determine which value of feedback factor β makes $A_{vf} = A_v$?

a zero
b less than unity
c unity
d greater than unity

Short answer questions

12.13

a Using a suitable block diagram describe how feedback is applied to an amplifier.
b Identify the function of each element in your diagram.
c Define the conditions necessary for positive and negative feedback.

12.14 Using the terms A_v = open loop gain
β = feedback factor
v_i = input voltage
v_s = signal voltage
v_o = output voltage
v_f = feedback voltage,
derive the general expression for the stage gain A_{vf} of a feedback amplifier.

12.15 Sketch and describe the feedback of a practical circuit employing;

a series current-derived feedback
b shunt voltage-derived feedback

13 Stabilised power supplies

This chapter describes the requirements for maintaining a constant voltage across a load. The aims of this chapter are as follows:

- to describe the main categories of power supplies and to provide an appreciation of their stablisition
- to describe the main elements of a stabilised power supply
- to describe and explain simple series and shunt stabilised power supply circuits.

13.1 Introduction

In electronics the term **stabilised power supply** usually describes an electronic circuit which is able to provide a known and controlled value of output voltage over a range of current.

Sometimes power supplies are just converters (say a.c. to d.c.). Stabilised power supplies are often converters but are always controllers: They maintain values of voltage (and sometimes current) at a particular level for a specified range of loads.

13.2 Main categories of power supplies

There are four main categories of power supplies;

a *a.c.* to *d.c.*
b *d.c.* to *d.c.* (one d.c. level to another)
c *a.c.* to *a.c.* (one a.c. level to another)
d *d.c.* to *a.c.*

In this text the first category is considered alone. Power supplies which provide a.c. to d.c. conversion comprise most of those in use in electronics. The simplest d.c. operates from the mains and comprises a transformer to reduce the a.c. voltage, a full-wave rectifier and a smoothing circuit (e.g. inductance/capacitance filter).

Such a basic supply is said to be **unstabilised** (or **unregulated**) because even though the a.c. ripple is reduced to an acceptable minimum, the output voltage still is affected by three factors; first, any a.c. supply amplitude variation is reflected in the d.c. output voltage; second, if the load current demand changes, the d.c. output voltage also changes; and third, a change in output current may result from a change in temperature, particularly if semiconductor devices are involved.

Many electronic circuits require a constant supply voltage for them to function correctly. With a varying supply voltage, the gain of an amplifier may change and the frequency of an oscillator may drift, particularly multivibrators where the switching time depends on the supply voltage (see Chapter 14).

13.3 Regulation and stability

Power supplies can be designed to overcome the disadvantages described in the previous section. They are called **stabilised** (or **regulated**) power supplies. Although the terms stabilised and regulated are often used interchangeably in this context, they refer to two different functions:

Regulation is a measure of how the output voltage remains constant for an increasing load current (Fig. 13.1a). A regulated supply delivers more or less current as demanded with little change in voltage.

Stability is a measure of how constant the output voltage remains for a given fluctuation of input voltage. A simple definition of the constancy of a power supply is given by the **stability factor** S where

$$S = \frac{\text{change in output voltage}}{\text{change in input voltage}} = \frac{\Delta V_o}{\Delta V_i} \qquad (13.1)$$

To make a variable supply more constant a stabilising circuit is required (Fig. 13.1b).

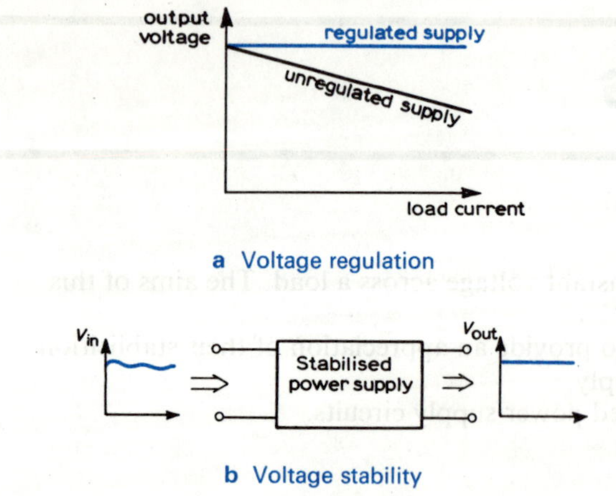

a Voltage regulation

b Voltage stability

Fig. 13.1

13.4 Elements of a stabilised power supply

Fig. 13.2 shows a block diagram of the basic elements of a stabilised power supply employing a **comparator technique**. The action of stabilising a power supply in this way is similar to the application of negative feedback in an amplifier. The main difference is that the output voltage is not fed back directly to the input, it is compared first with a voltage reference source by a **comparator** which sends a feedback signal to the **control element**. Obeying the signal from the comparator, the control element compensates for any difference between the true output voltage and the desired output voltage.

The individual elements are discussed in more detail as follows:

The **unstabilised voltage source** is generally the output of a full-wave rectifier circuit with smoothing.

The **control element** is used to compensate for any variation in the unstabilised supply. This device is often a transistor or sometimes for high voltages a valve, although advances in semi-conductor technology now make the use of valves rare.

The control element acts as a buffer between the unstabilised supply and the output, where the output is generally lower than the input, i.e. a voltage is dropped across the device. If the input is say 20 V, typically 5 V to 10 V would be dropped across the control element. In this way a reservoir voltage is available to compensate for any tendency to fluctuate positively or negatively. If the output voltage was the same as the input, only positive fluctuations could be corrected.

The comparator compares the actual output voltage with the desired output voltage obtained from a reference source. A signal proportional to the difference between the output and reference voltages is passed to the control element which enables compensating action to take place.

The **reference source** is often a zener diode. The zener diode is itself the simplest form of voltage stabilisation, but the maximum output current has to be limited to the maximum rating of the diode (see Section 2.12). The maximum current rating of the diode alone is often inadequate.

Many supplies include a **protection circuit** which reduces the output current to a safe level should the output be accidentally short-circuited. The switching to a safe current level has to be done within a fraction of a second to prevent permanent damage to the power supply or the load it is feeding.

Power supplies can be designed with the control element in series or in shunt with the load.

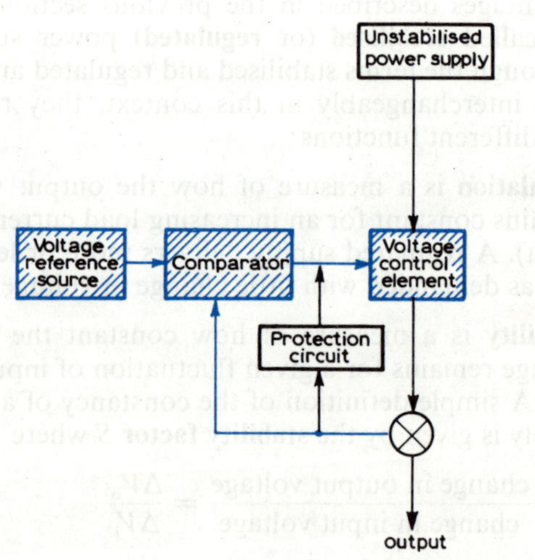

Fig. 13.2 Block diagram of a stabilised power supply

13.5 Series stabilised power supply

A series stabilised power supply is shown in Fig. 13.3a. The basic elements identified in the block diagram of Fig. 13.2 are as follows.

The **series control element** is the transistor, which is held in the conducting state by the constant reference voltage V_z standing at the base. Because the transistor base current is fixed by the base voltage, the amplified current through the collector and emitter also is steady, giving a constant output voltage across the load resistor R_L. Current is therefore supplied through the path which includes R_s and the transistor itself.

The difference between V_o and V_i is the p.d. across R_s and the collector-emitter p.d. V_{CE}. This value is typically 5 V to 10 V and acts as the voltage reservoir to absorb any fluctuations in V_i.

Part of the transistor also serves as the comparator. The difference between V_o and V_z is the base-emitter potential difference V_{BE}. If the value of V_o tends to change because of varying load conditions, V_{BE} changes. This has the effect of automatically adjusting the conduction through the transistor and V_{CE} changes to compensate.

The **voltage reference source** is provided by the zener diode (see section 2.12). R_B limits the current through the diode.

An alternative voltage reference configuration involves connecting a resistance potentiometer across the zener diode (Fig. 13.3a insert). The voltage to the base of the transistor is taken from the slider which enables the voltage reference to be varied. Therefore the output voltage can be set manually.

Short-circuit protection is provided by the series resistor R_S, which limits the current to a safe maximum.

Summary: The action of a series stabilised power supply can be summarised as follows. If the load current varies, the change in demand for current is accommodated by drawing more or less from the supply input, but because of the transistor action, there is no change in V_o. If the unstabilised supply fluctuates, the change in p.d. is developed across R_S and R_B. Again, V_o is relatively unaffected.

13.6 Shunt stabilised power supply

The circuit diagram of a simple shunt stabilised power supply is shown in Fig. 13.3b.

The bias voltage at the base of the shunt transistor is held constant by the reference voltage V_z. If the unstabilised supply voltage should increase, the zener diode conducts more current, turning on the transistor harder which also conducts more current through the collector-emitter path. Thus the current handling capacity of this circuit is much greater than that of the zener diode alone. This compensation of current enables the current supplied to a constant load to be held steady. Any fluctuation of V_i is developed across R_S which also serves as short-circuit protection.

If the load changes, demanding more or less current, the conduction through the transistor changes to accommodate this whilst the output voltage V_{CE} is held constant.

Review

Check your knowledge and understanding. You should:

a Be able to identify the requirements for maintaining a constant voltage output across a load.

b Know how to sketch a block diagram of a series stabilised power supply using the comparator technique.

c Understand the operation of the power supply of **b**.

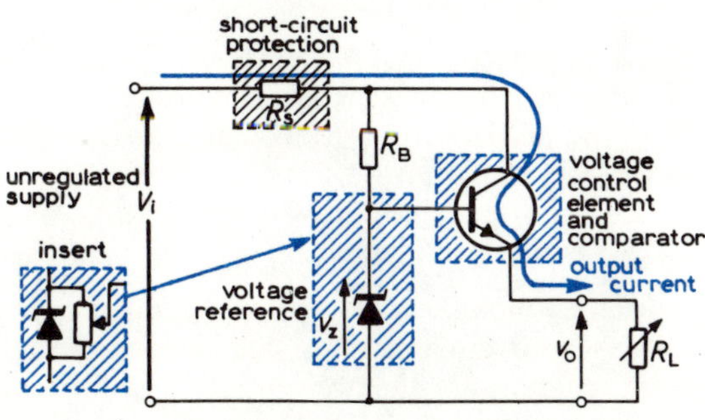

a Series stabilised supply

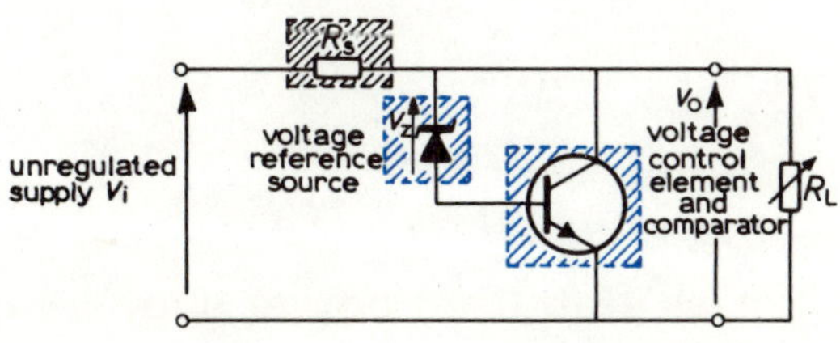

b Shunt stabilised supply

Fig. 13.3

Exercise

13.1 When the output voltage is constant over a range of loads identify if the power supply is:

a regulated or **b** stabilised

13.2 Identify which of the following statements about stabilised power supplies are true and which are false.

- **a** The comparator compares the output voltage with the input voltage.
- **b** The output voltage of a stabilised power supply is generally lower than the unstabilised input voltage.
- **c** The unstabilised voltage has a higher stability factor than a stabilised voltage.
- **d** The control element is used to compensate for any variation in the unstabilised supply.

Self assessment questions

The following are typical assessment questions.

Multiple choice questions

13.3 When the output voltage is constant for a varying input voltage, identify if the power supply is regulated or stabilised.

13.4 Identify which of the following statements about stabilised power supplies are true and which are false:

- **a** The comparator compares the output voltage with a reference source.
- **b** The reference source is usually the unstabilised supply.
- **c** The control element is switched into the circuit by the comparator only when the input voltage exceeds the reference voltage.
- **d** The output voltage of a stabilised power supply is generally lower than the unstabilised voltage input.

Short answer questions

13.5 Sketch a block diagram of a series stabilised supply, labelling the sections

- **a** control element
- **b** comparator
- **c** voltage reference source

13.6 With reference to your block diagram of Question 13.5 describe the basic operation of the stabilised power supply.

13.7 State three of the main factors which affect the output voltage of an unstabilised power supply.

14 Pulse generators and pulse-shaping networks

This chapter describes electronic pulse generators and special circuits for obtaining appropriate pulses of the required shape and amplitude for given applications. The aims of this chapter are as follows:

- to describe and provide examples of the main types of pulse generators (astable, bistable and monostable)
- to describe and provide an appreciation of pulse shaping circuits.

14.1 Multivibrators

Rectangular waveforms and pulses have many applications in electronics ranging from controlling the operation of a computer to pulse testing the response of an amplifier as outlined in Section 7.1.

For these purposes three types of rectangular outputs are produced by related pulse generators called **multivibrators.**

Astable: A rectangular waveform with repeated cycles is generated by an astable multivibrator (Fig. 14.1a).

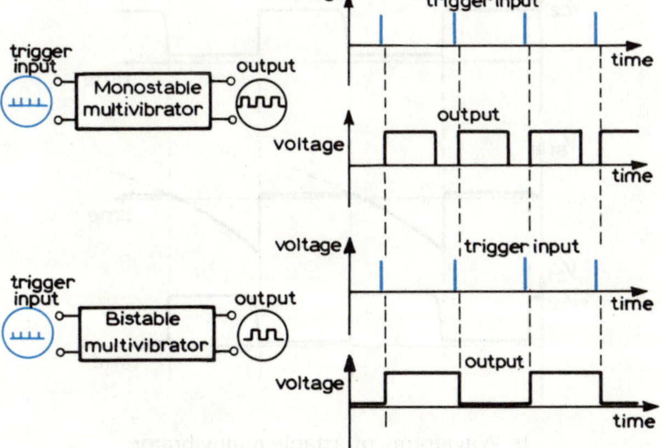

Fig. 14.1 The three types of multivibrator

The astable multivibrator is said to be a free-running oscillator because it requires no external control.

Monostable: A single pulse of constant amplitude and fixed time-duration is generated by a monostable multivibrator (Fig. 14.1b). The output pulse is started by a short-duration trigger pulse which may be of variable amplitude and of no fixed time-duration.

Bistable: A third type provides a change of voltage state (from high to low or low to high). This is a bistable multivibrator (Fig. 14.1c) which responds to the input of a trigger pulse.

Multivibrators are sometimes called **relaxation oscillators** because of their sudden change of state or relaxation of circuit conditions.

14.2 Requirements of transistor multivibrators

A multivibrator requires circuit elements which switch from one state to another. This can be done by using two transistor common emitter amplifiers, the output of each feeding the input of the other through coupling networks (Fig. 14.2a).

The usual arrangement in diagrammatic form is shown in Fig. 14.2b where transistor T_1 and R_{L1} form the basic common-emitter configuration of amplifier 1 and transistor T_2 and R_{L2} form the common-emitter configuration of amplifier 2.

The cross-coupling ensures that as one transistor is ON, the other is OFF and a switching of states between the two is possible. Each transistor is biased

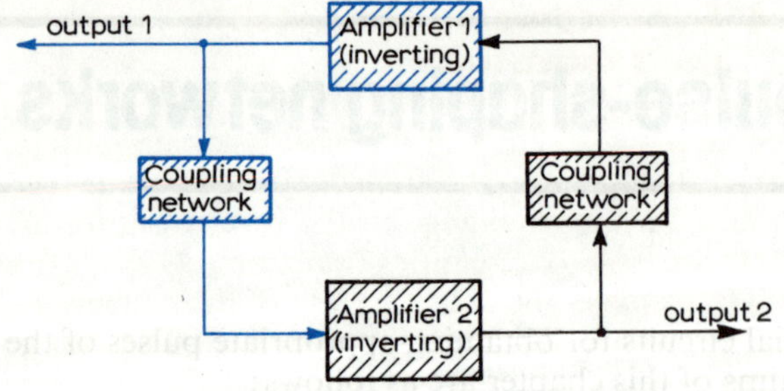

a Block diagram of multivibrator

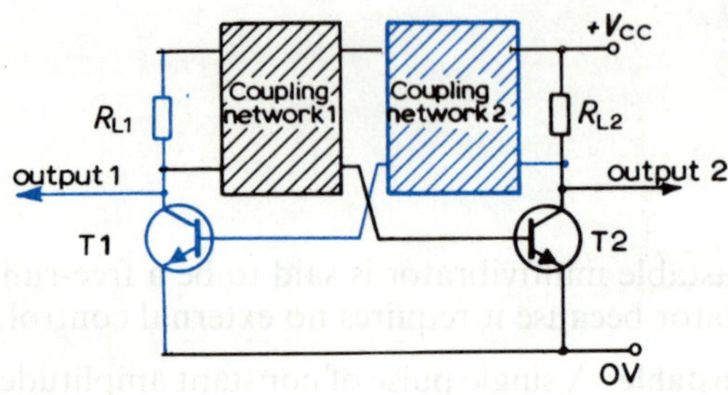

b Basic transistor multivibrator

Fig. 14.2

to operate as a switch (see Section 8.7) where saturation is required of the transistor in the ON state.

Whether the multivibrator operates as astable, monostable or bistable depends on the components used in the coupling circuits.

Multivibrators are two inverting amplifiers providing each other with 100% positive feedback.

14.3 Astable multivibrator

The astable multivibrator requires symmetrical cross-coupling networks each comprising one capacitor and one resistor (Fig. 14.3a). The sequencing of the collector output waveforms V_{C1} and V_{C2} and the base waveforms V_{B1} and V_{B2} are shown in Fig. 14.3b.

Immediately after the supply V_{CC} is switched on (at $t = t_0$ on the waveforms of Fig. 14.3b) voltages V_{C1} and V_{C2} appear at the collectors of T_1 and T_2:

$$V_{C1} = V_{CC} - I_{C1}R_{L1} \quad (14.1)$$

$$\text{and} \quad V_{C2} = V_{CC} - I_{C2}R_{L2} \quad (14.2)$$

The values of V_{C1} and V_{C2} are almost equal to $+V_{CC}$ because no collector current has yet had time to flow, therefore no p.d. is dropped across R_{L1} and R_{L2}.

Current flows into the bases of both transistors through their respective base resistors R_{B1} and R_{B2}. (One transistor turns ON before the other because their characteristics differ slightly, even if they were manufactured in the same batch.)

Assume that T_1 is the first to turn ON and follow the cycle of events through points 1 to 4 as shown above the waveforms in Fig. 14.3b.

Point 1 When the base-emitter voltage V_{B1} exceeds the base-emitter barrier potential, T_1 turns ON and current flows in the collector of T_1. Because of the resulting voltage drop across R_{L1}, the value of V_{C1} drops to slightly above zero volts.

The p.d. appearing across C_2 is initially almost the whole of the supply voltage (because $V_{C1} \approx V_{CC}$ at one side of C_2 and the base of T_2 is zero volts at the other).

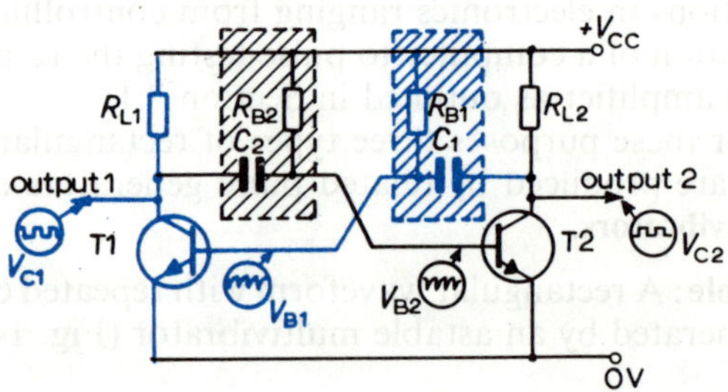

a The astable multivibrator circuit

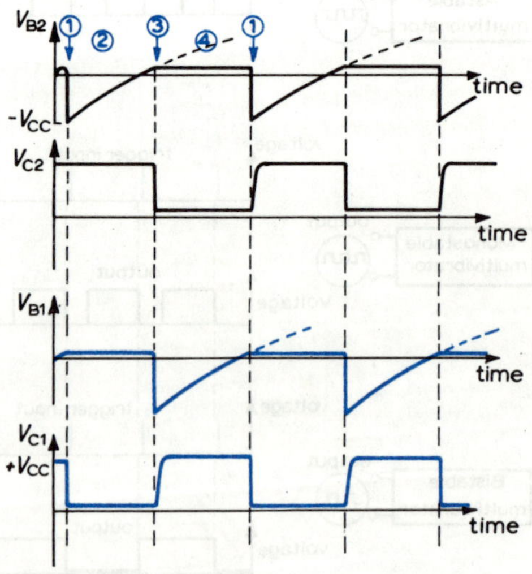

b Waveforms of astable multivibrator

Fig. 14.3

An important aspect of the operation of multivibrators is that the p.d. across a capacitor cannot change instantaneously. This is because charge requires a finite time to flow in or out of a capacitor. If the voltage at one side of a capacitor suddenly changes from positive to zero, the voltage at the other side falls from zero to a negative voltage to maintain the same p.d. (Fig. 14.4). Bearing this in mind, and returning to the waveforms of Fig. 14.3b: As soon as V_{C1} drops to zero, the voltage V_{B2} falls to a value $-V_{CC}$ because of the potential across C_2. This reverse biases the base-emitter junction of T_2, holding it OFF. V_{C2} is therefore almost equal to V_{CC}.

Point 2 Capacitor C_2 begins to charge towards the value of $+V_{CC}$ through R_{B2}. The rate of charging is governed by the C_2R_{B2} time constant.

Point 3 When V_{B2} goes positive (by about 0.6 V for silicon) T_2 is turned ON. As V_{C2} drops to almost zero, V_{B1} falls to almost $-V_{CC}$ because of the p.d. across C_1. Therefore T_1 is turned OFF.

Point 4 As C_1 charges towards V_{CC} through R_{B1}, V_{B1} rises until it is sufficient to turn ON T_1.

The whole cycle is then repeated for as long as the supply is connected. A rectangular waveform can be tapped from the collector of T_1. A similar waveform in antiphase is available from the collector of T_2.

The frequency of oscillation is determined by the $R_{B1}C_1$ and $R_{B2}C_2$ time constants. A large time constant gives a longer periodic time and therefore a lower frequency. Short time constants give high frequencies.

The example described here has a 1:1 **mark-space ratio** because the two time-constants are equal. Different time-constants for each coupling network can be used to change the mark-space ratio of the output waveform. For example, if $R_{B2}C_2$ had twice the value of $R_{B1}C_1$ the mark-space ratio would be 2:1.

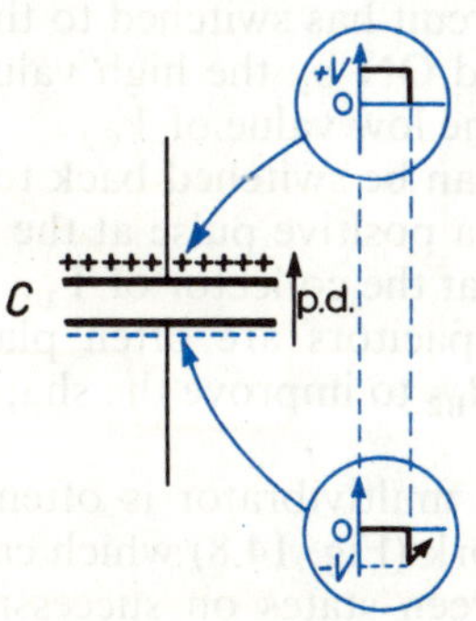

Fig. 14.4 PD across capacitor

14.4 Synchronising the astable multivibrator

The frequency stability of the basic astable multivibrator circuit is low. The frequency varies with temperature because a change in the reverse saturation current of the base-emitter junction of the OFF transistor affects the charging rate of the capacitor.

To overcome this instability, the switching can be synchronised using short-duration pulses generated by a frequency stable oscillator (Fig. 14.5a).

The synchronising pulses are fed to the base of T_1 and T_2 through diodes to prevent the bases becoming electrically connected together (Fig. 14.5b). When a pulse takes the base voltage of the OFF transistor above zero, the circuit is switched.

Notice that in the example shown in Fig. 14.5, only

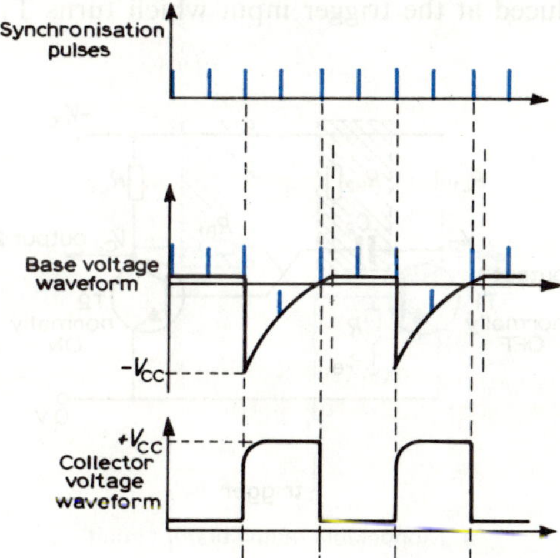

a Synchronising the astable multivibrator with synchronisation pulses

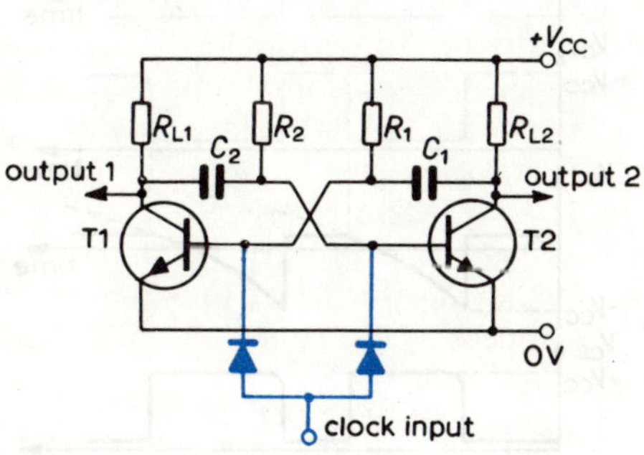

b Trigger circuit for astable multivibrator

Fig. 14.5

every second trigger pulse switches the circuit. In this way the multivibrator serves as **divide-by-two circuit**. By careful choice of frequencies up to divide-by-ten is possible.

14.5 Monostable multivibrator

The basic circuit of the monostable multivibrator is shown in Fig. 14.6a. The coupling of T_2 is an *RC* charging combination. The coupling at the base of T_1 is a single resistor R_{B1} connected to the collector of T_2. An additional negative $-V_B$ at the base of T_1 ensures that it is normally OFF. The high value of V_{C1} normally holds T_2 ON.

This situation prevails until a positive pulse is introduced at the trigger input which turns T_1 ON.

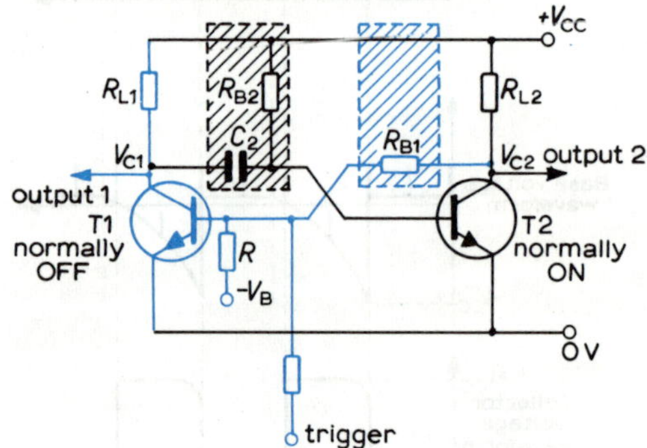

a Monostable multivibrator circuit

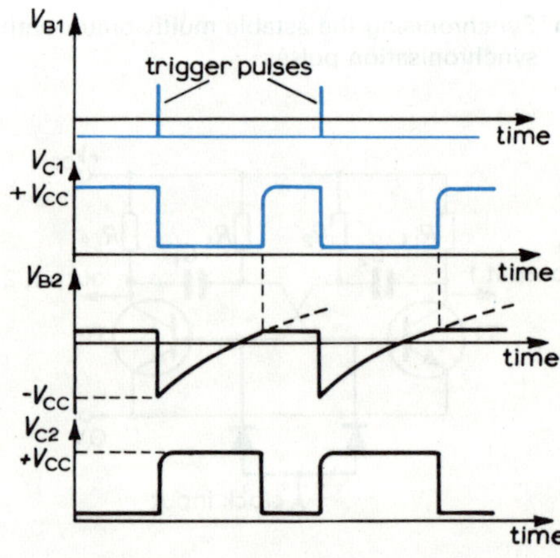

b Waveforms of monostable multivibrator

Fig. 14.6

As collector current flows through R_{L1}, the value of V_{C1} falls to almost zero (Fig. 14.6b) Because the p.d. across C_2 cannot change instantaneously, V_{B2} falls to almost $-V_{CC}$. Therefore T_2 is held OFF until V_{B2} rises to the value which turns T_2 ON. The circuit has then returned to its original state.

The duration of the pulse appearing at output 2 depends upon the $R_{B2}C_2$ time constant. A longer time-constant would give a slower charging rate and hence a longer pulse duration.

A negative pulse at the base of T_2 has the same switching effect. Consequently the monostable multivibrator can be switched temporarily to an unstable state (sometimes referred to as **quasi-stable** or **apparently stable**) and always reverts to a stable state after a fixed duration of time.

A **speed-up capacitor** is often connected across R_{B1} to achieve speedier switching. Because the p.d. across the capacitor cannot change instantaneously, it has the effect of immediately transferring the positive potential of the trigger pulse to the collector of T_2. This improves the sharpness of the pulse.

14.6 Bistable multivibrator

The bistable multivibrator (Fig. 14.7) has two stable states between which the circuit can be controlled under the control of trigger pulses.

When the circuit is first connected to the supply, one transistor will be turned ON before the other. Assume that T_1 is ON. V_{C1} is low so T_2 is held OFF. Because no collector current flows through R_{L2}, V_{C2} is high, ensuring that T_1 is held ON.

This stable condition prevails until the circuit is switched in one of two ways. A positive pulse on the base of T_2 turns it ON, or a negative pulse at the collector of T_2 turns T_1 OFF.

When the circuit has switched to the second stable state, T_2 is held ON by the high value of V_{C1}. T_1 is held OFF by the low value of V_{C2}.

The circuit can be switched back to the first stable by either a positive pulse at the base of T_1 or a negative pulse at the collector of T_1.

Speed-up capacitors are often placed in parallel with R_{B1} and R_{B2} to improve the shape of the switching waveform.

The bistable multivibrator is often provided with a trigger network (Fig. 14.8) which causes the circuit to switch between states on successive pulses. This is often called a **steering circuit** because the incoming pulse is steered to the base of the transistor which is

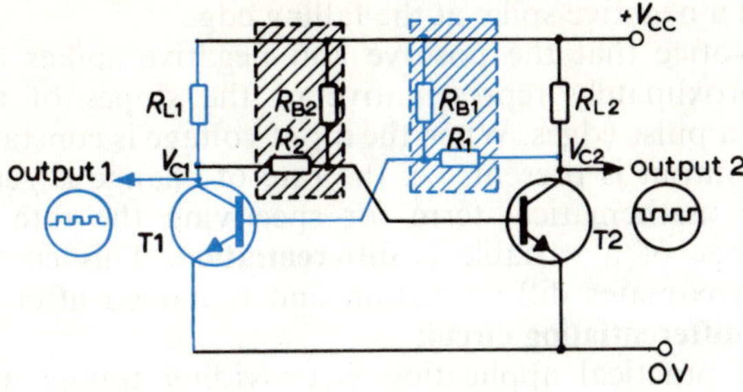

a Bistable multivibrator circuit

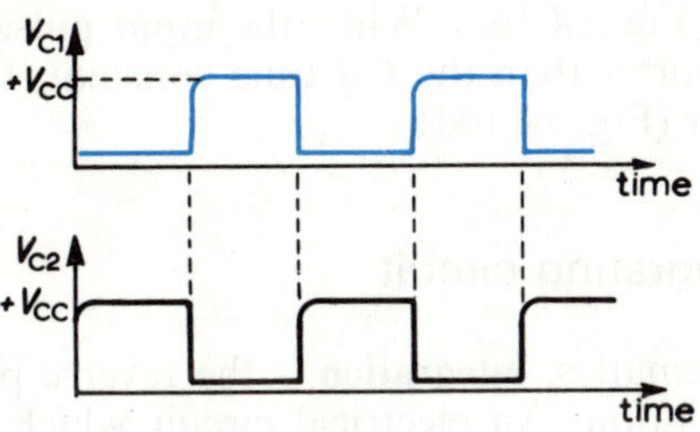

b Waveforms of bistable multivibrator

Fig. 14.7

OFF. A reset facility to return the circuit to its first stable state also is provided.

Because the circuit is switched between states at every trigger pulse, any output shows one pulse for every two trigger pulses. The output frequency is therefore half the trigger frequency. In this way the bistable is often used as a **divide-by-two** (or a **scale-of-two**) element.

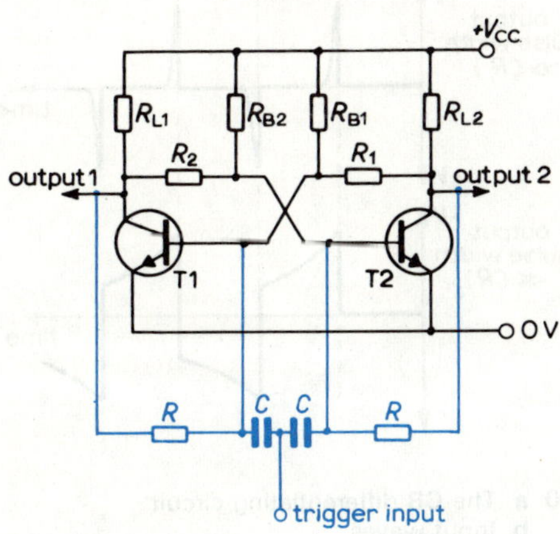

Fig. 14.8 Steering network for bistable multivibrator

Review

Check your knowledge and understanding. You should:

a Be able to state the requirements of a transistor multivibrator.
b Know the difference between the three different types of multivibrator; astable, monostable, and bistable.
c Understand the action of the three different types of multivibrator by deriving waveforms present around the circuits.
d Understand the need for synchronising and triggering multivibrators.
e Know the methods of synchronising and triggering multivibrators.

Exercise

14.1 Identify the types of multivibrator.
a The circuit is switched between two stable states by the application of a trigger pulse.
b The circuit is temporarily switched from a stable state to an unstable state by the application of a trigger pulse.
c The circuit is switched between two unstable states under the control of *CR* timing networks.

14.2 Identify which of the following are true or false.
a Each transistor amplifying element of a multivibrator is provided with 100% negative feedback.
b Each transistor is saturated when on.
c A phase shift of 180° is required by each amplifying element.
d The output from one collector is in phase with the output from the other.
e Speed up capacitors increase the frequency of the astable multivibrator.

14.3 Identify which of the following will trigger any *npn* transistor multivibrator.
a A negative pulse at the collector of the ON transistor
b A positive pulse at the base of the OFF transistor
c A negative pulse at the base of the OFF transistor
d A positive pulse at the base of the ON transistor

14.4 An astable multivibrator has the following *CR* combinations:
Coupling network 1: $C = 0.1\ \mu\text{F}$, $R = 1\ \text{k}\Omega$
Coupling network 2: $C = 0.2\ \mu\text{F}$, $R = 2\ \text{k}\Omega$
Determine the mark-space ratio.

14.7 Real pulses

So far, it has been assumed that rectangular pulses have vertical sides and a flat top. In practice they can approach this but do not quite attain this ideal shape.

A practical pulse, exaggerated for clarity is shown in Fig. 14.9. The following terms are used to describe a real pulse:

Pulse amplitude is the peak or maximum value of the pulse.

Rise time t_R is the time taken for the pulse to rise from 10% to 90% of its peak value. The reason for specifying lower and upper limits is that real pulses deviate from linearity near the bottom and top.

Fall (or **decay time**) t_F is the time taken for the pulse to fall from 90% to 10% of its peak value.

Pulse duration (or **pulse width**) is the duration of time when the pulse is over 90% of its peak value.

These terms are important when considering particular applications for which the pulse is used. For example, the controlling of events (clocking) of digital equipment often requires a minimum and maximum pulse amplitude, pulse width and clearly defined rise and fall times. There are stringent requirements covering the rectangular qualities of timing pulses in televisions.

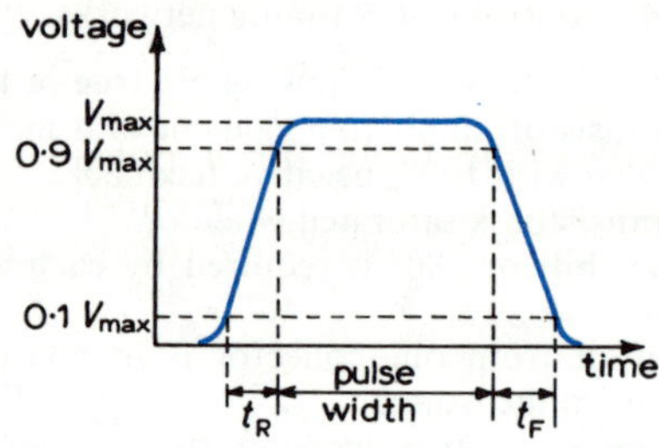

Fig. 14.9 Pulse characteristics

14.8 Pulse-shaping circuits

Two simple pulse-shaping circuits are commonly used to modify rectangular waveforms. These employ a resistor and capacitor and are called differentiating and integrating networks.

CR differentiating circuit

The combination of a series capacitor and shunt resistor (Fig. 14.10a) produces an output which is proportional to the slope or rate of change of the input.

When the input is a rectangular pulse, the output is a short duration positive spike at the rising edge and a negative spike at the falling edge.

Notice that the positive and negative spikes are approximately representative of the slopes of the input pulse edges. When the input voltage is constant, no output is present, i.e. the rate of change is zero. The mathematical term for specifying the rate of change of a variable is **differentiation**. This circuit approximates differentiation and is named after it; the **differentiating circuit**.

A practical application is providing timing and trigger pulses for electronic circuits.

When the input pulse width is much greater than the *CR* time constant, the output spikes are very narrow (Fig. 14.10c). When the input pulse width is much shorter than the *CR* time constant, the spikes are fatter (Fig. 14.10d).

CR integrating circuit

In mathematics, **integration** is the reverse process of differentiation. An electrical circuit which approximately performs this operation comprises of a resistor

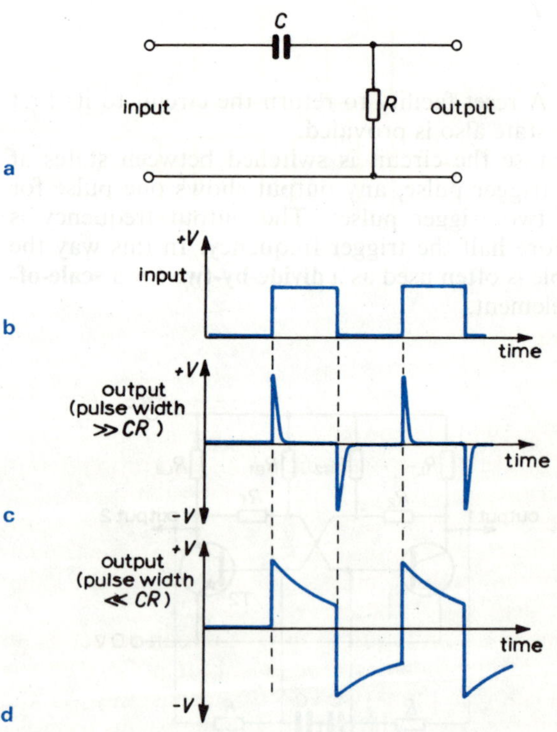

Fig. 14.10 a The CR differentiating circuit
b Input waves
c Output when pulse width ≫ CR
d Output when pulse width ≪ CR

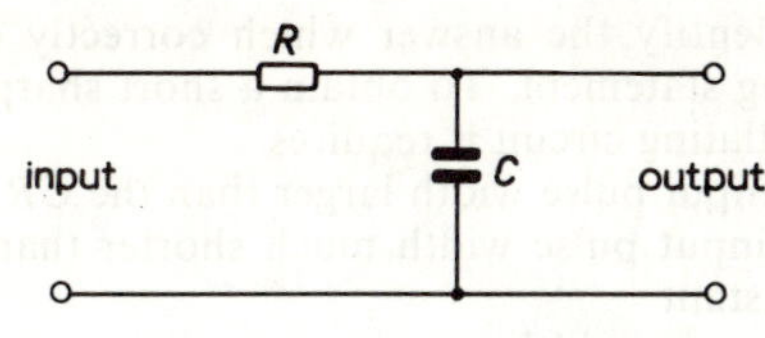

CR integrating circuit

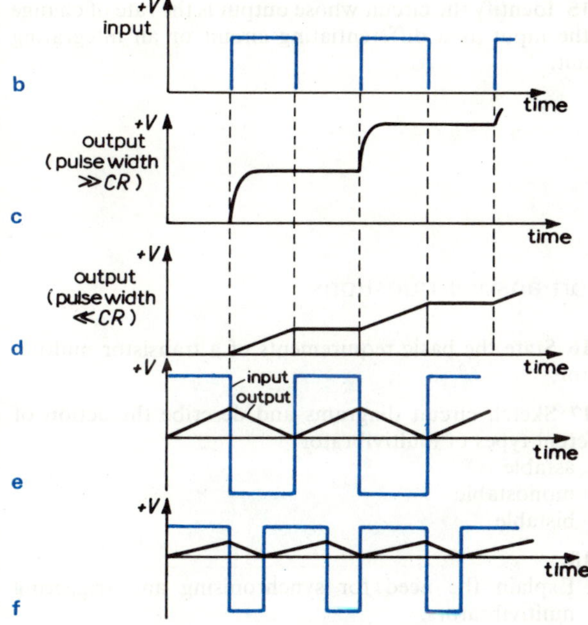

Input and output waveforms

Fig. 14.11

in series with the input and a capacitor in shunt (Fig. 14.11a). The circuit is called an **integrating circuit**.

The output resulting from a rectangular input pulse depends on the *CR* time constant. If the pulse width is much larger than the *CR* time constant, the capacitor charges and the output voltage rises to the value of the peak input voltage (Fig. 14.11c).

The condition required for the circuit to act as an electrical integrator is that the input pulse width is much shorter than the *CR* time constant. As the capacitor charges, the output voltage begins to move towards the input pulse peak voltage, but does not have time to move beyond the lower part of the charging curve (Fig. 14.11d) before the pulse ends. Here the output can be considered linear.

If the capacitor cannot discharge between pulses, a series of positive pulses gives a sloping stepped output (Fig. 14.11d). When the input pulse goes negative (Fig. 14.11e) *C* discharges back through *R*. By introducing different positive and negative pulse amplitudes and a different mark-space ratio (Fig. 14.11f) the output can be made a ramp waveform with any specified sweep and flyback.

Because the output from this basic integrating circuit is small, amplification is sometimes necessary (see Section 15.11).

Review

Check your knowledge and understanding. You should:

a Know how to sketch and label a rectangular pulse waveform showing pulse width, pulse amplitude, rise-time, and fall-time.

b Know the diagram of a differentiating circuit.

c Understand the form of output waveform of a differentiating circuit for a rectangular pulse much longer, and a rectangular pulse much shorter than the *CR* time.

d Know the diagram of an integrating circuit.

e Understand the form of output waveform of an integrating circuit when a rectangular pulse is applied to the input for a pulse width much greater, and a pulse width much smaller than the *CR* time.

Exercise

14.5 Identify the answer which correctly completes the following statement. The rise time of a rectangular pulse is defined as

a the time taken to rise from 5% to 95% of its peak value

b the time duration of the maximum amplitude

c the time taken to rise from 10% to 90% of its peak value

d the time duration between the 90% points of the maximum amplitude

14.6 Identify the differentiating circuit components

a a shunt *C* with series *R*

b a series *C* with shunt *R*

14.7 Identify the answer which correctly completes the following statement. To obtain a sawtooth waveform from a *CR* integrating circuit

a the input pulse width should be much greater than the *CR* time constant

b the input pulse width should be much smaller than the *CR* time constant

c the *CR* time constant is irrelevant

d the input should be sinusoidal.

Self assessment questions

The following are typical of questions you will be asked in an assessment test.

Multiple choice questions

14.8 Identify which of the following is the main requirement of a transistor multivibrator;
- **a** two transistor amplifiers with 100% negative feedback between them
- **b** two transistor amplifiers, each amplifying alternate half cycles
- **c** two transistor amplifiers with common outputs
- **d** two transistor amplifiers with 100% positive feedback between them

14.9 Fig. 14.2b shows the basic multivibrator circuit connected by cross-coupling networks. Identify the type of multivibrator by inserting the following components in the cross-coupling networks;
- **a** two resistors in each network
- **b** a resistor and capacitor in each network
- **c** a resistor and capacitor in one network and a resistor in the other

14.10 Identify how the pulse frequency of an astable multivibrator is determined.
- **a** by gain of each transistor
- **b** by the RC combinations
- **c** by the supply voltage V_{CC}
- **d** by the collector load resistors.

14.11 Identify the answer which correctly completes the following statement. The base waveform of either transistor in an astable multivibrator is
- **a** a rectangular wave in phase with the output
- **b** a series of positive spikes slowly discharging
- **c** a rectangular wave 180° out of phase with the output
- **d** a series of negative spikes, slowly charging

14.12 Identify the answer which correctly completes the following statement. If the capacitor in a monostable multivibrator was made larger the pulse duration would
- **a** increase
- **b** decrease
- **c** remain the same
- **d** become zero because the circuit would act as a bistable.

14.13 Identify the answer which correctly completes the following statement. To obtain a short sharp pulse from a differentiating circuit it requires
- **a** an input pulse width larger than the *CR* time constant
- **b** an input pulse width much shorter than the *CR* time constant
- **c** any pulse width
- **d** a sinusoidal input

14.14 Identify the components of an integrating circuit
- **a** a shunt capacitor with series resistor
- **b** a series capacitor with a shunt resistor

14.15 Identify the circuit whose output is the rate of change of the input as a differentiating circuit or an integrating circuit.

Short answer questions

14.16 State the basic requirements of a transistor multivibrator.

14.17 Sketch circuit diagrams and describe the action of different types of multivibrator
- **a** astable
- **b** monostable
- **c** bistable

14.18
- **a** Explain the need for synchronising and triggering multivibrators.
- **b** State the methods of applying trigger pulses to
  - **i** a monostable
  - **ii** an astable
- **c** Describe the function of a steering network in a bistable multivibrator

14.19 Sketch and label a rectangular pulse showing pulse amplitude, pulse width, rise-time and decay-time.

14.20 Sketch the diagram of an integrating circuit and output waveform when
- **a** the pulse width is much greater than the *CR* time
- **b** the pulse width is much smaller than the *CR* time

15 Integrated circuits

This chapter describes the manufacture, advantages and examples of integrated circuits (ICs). The aims of this chapter are as follows:

- to explain the manufacture of integrated circuits and how they are packaged
- to provide an appreciation of the advantages of ICs
- to describe various types of IC in common usage.

15.1 Introduction

Before 1960 all commercial electronic circuits were constructed of discrete components (each component was manufactured individually and then soldered together on to a circuit board).

In the early 1960s semiconductor manufacturers began to employ a technique whereby more than one active component could be made side by side in a single piece called a **chip**. Methods were developed to include passive components alongside active components within the same chip so that complete circuits were integrated into one piece of material. An entire circuit (e.g. amplifier, multivibrator, arrays of logic gates and so on), consisting of transistors, diodes, capacitors and resistors could be manufactured and interconnected in a single semiconductor device (Fig. 15.1). Today, these complex semiconductor devices, called **integrated circuits (ICs)**, replace circuits which require many discrete components.

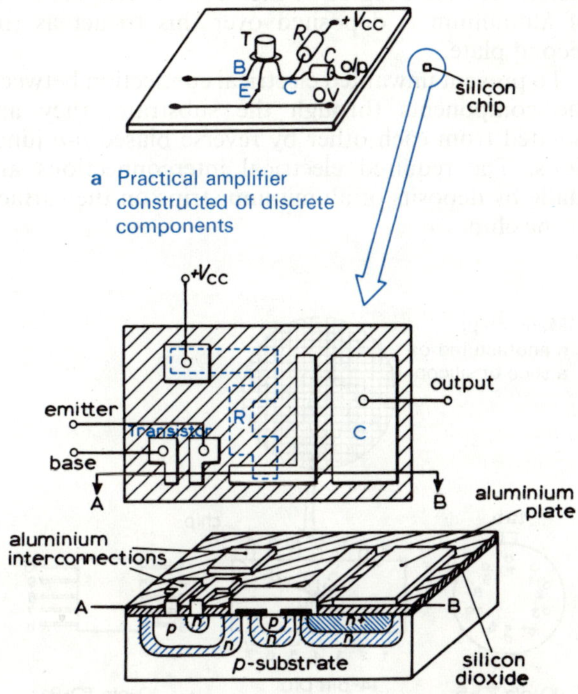

Fig. 15.1 Monolithic IC compared with discrete components

15.2 Monolithic integrated circuits

Monolithic ICs are produced by manufacturing the many active devices and passive components of a circuit within a single chip of semiconductor. (*Mono* means single, and *lithic* means stone.) Silicon is by far the most common semiconductor material used in the manufacture of ICs.

It is possible to manufacture bipolar transistor ICs and MOSFET ICs. Because of its superior operating qualities, the bipolar transistor is used in many digital ICs and for IC amplifiers and related circuits. The MOSFET IC, however, requires as few as one-fifth of the processing steps than is required for the bipolar IC. MOS circuits are cheap because of their simplicity. Their main advantage lies in the possibility of thousands of logic elements being included in a single array.

15.3 Manufacture of monolithic ICs

The monolithic IC is manufactured using **planar technology**, meaning that the semiconductor is processed in layers or planes. All the components are formed by a series of dopant diffusions into the basic semiconductor chip known as a **substrate**. Fig. 15.1 shows a diagram of a monlithic IC.

Transistors and diodes are the smallest devices in an IC, formed by successive diffusions of acceptors and donors to form layers of *p*- and *n*- type regions.

Resistors in monolithic ICs consist of a thin strip of semiconductor with low conductivity, zig-zagged to increase the resistance path length for a given chip area. As they take up quite a lot of room their maximum value is limited to 20 kΩ.

Capacitors in monolithic ICs are formed by depositing a layer of silicon dioxide, which acts as a dielectric, over a region of high conductivity. A layer of aluminium is deposited over this to act as the second plate.

To prevent unwanted electrical connection between the components through the substrate, they are isolated from each other by reverse biased *p-n* junctions. The required electrical interconnections are made by deposits of aluminium strips on the surface of the chip.

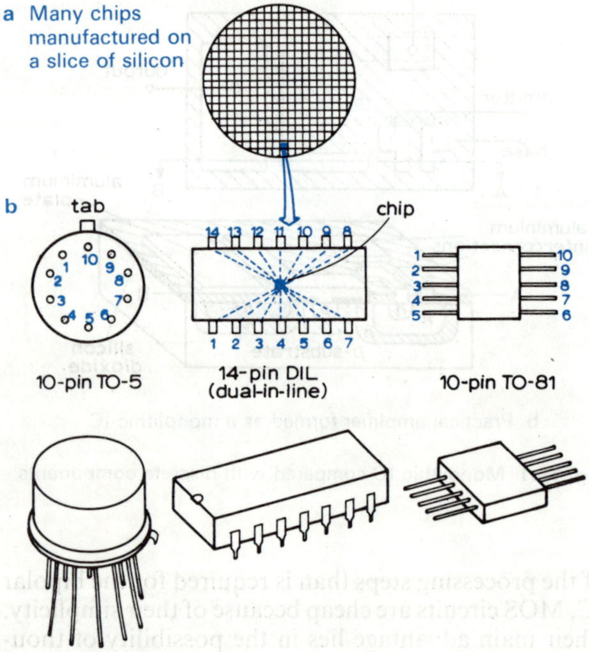

a Many chips manufactured on a slice of silicon

b 10-pin TO-5 · 14-pin DIL (dual-in-line) · 10-pin TO-81

Fig. 15.2 IC packaging

Because of its relatively large size, the inductor is rarely included in an IC. If required, it is connected externally.

15.4 Integrated circuit package types

Many identical ICs can be manufactured simultaneously on a thin disc of silicon (Fig. 15.2a). The disc is then cut and each chip packaged separately.

Each IC chip requires the connection of terminals for electrical access and then is sealed in a plastic or ceramic package for protection. ICs are available in a variety of package types. A selection of those most commonly used are shown in Fig. 15.2.

The terminals are shown numbered for identification. Connections are required for power supply, inputs and outputs. Some connections are required for the facilities provided by individual ICs. For example, the performance of an IC amplifier may be changed in a desired manner by the external connection of a few discrete components to specified terminals.

Fig. 15.2 illustrates that the size of the packaging is much greater than the chip itself.

15.5 Advantages of integration

The advantages of integration are emphasised by their widespread use in industrial, military and domestic electronic equipment. The main advantages are:

Size: The functional area of an IC is typically a few square millimetres. Equivalent discrete circuits could be many thousands of times larger. Miniaturisation is important in computer and military applications.

Reliability: Because of the intimate contact between components within the semiconductor chip, connecting wires are eliminated. This reduces the problems of stray capacitance, inductance and the high unreliability associated with the wiring of discrete circuits. Unless misused, faults within the chips are rare.

Cost: A prime consideration of any manufacturer of electronic equipment is the cost of components. Today many ICs are available at a price comparable to that of discrete components. **Large scale integration** (LSI) involving many hundreds or thousands of components per chip and mass production help keep the cost to a minimum.

The speed of operation and low power consumption of many IC logic circuits are additional advantages.

15.6 Digital integrated circuits

Digital ICs are used with two-state logic signals which are often in the form of pulse trains. Many logic functions are available. For example, an IC catalogue identifies chips under the IC 74 series and the CMOS 4000 series which provide the logic functions in Chapter 8. However, there are usually many gates on one chip.

15.7 Linear integrated circuits

Linear ICs deal with analogue signals. Many of the circuits described in previous chapters are available as linear ICs or can be built around ICs with the addition of a few discrete components. These low cost chips enable the technician to build sophisticated electronic systems which would not be feasible if built entirely of discrete components.

Examples if linear IC applications are

- high gain direct coupled amplifiers
- differential amplifiers which amplify the difference between signals presented to two separate inputs
- audio amplifiers which are often supplied in pairs for stereophonic applications
- radio frequency and intermediate frequency amplifiers
- wideband amplifiers
- oscillators
- stabilised power supplies

Review

Check your knowledge and understanding. You should:

a Know the available range of linear integrated circuits.

b Appreciate the advantages of linear integrated circuits.

Exercise

15.1 Identify why a monolithic IC is so called. Because;
 a of its shape
 b it has only one terminal
 c it is manufactured as a single piece of silicon.
 d it contains one active device only.

15.2 Identify why linear ICs are so called. Because;
 a their connecting pins are in-line
 b they are manufactured as bipolar transistor only
 c they respond to an input signal in a linear manner
 d they are manufactured in bulk on a production line

15.3 Identify the package types described here:
 a two lines of connecting pins arranged either side of a long slender encapsulation
 b a circular arrangement of terminals

15.8 Operational amplifiers

Operational amplifiers are high gain direct-coupled amplifiers with very high input impedance and low output impedance. They are available as cheap ICs. A selection of standard types are the 741, 709, 710 and so on, all of which are made to their individual performance specifications by numerous manufacturers.

The general circuit symbol of the operational amplifier is shown in Fig. 15.3a and two standard IC packages illustrated in Fig. 15.3b.

The operational amplifier has two inputs, one an **inverting input (−)**, the other a **non-inverting input (+)**, and a single output. Signals presented to the inverting input appear at the output 180° out of phase with the

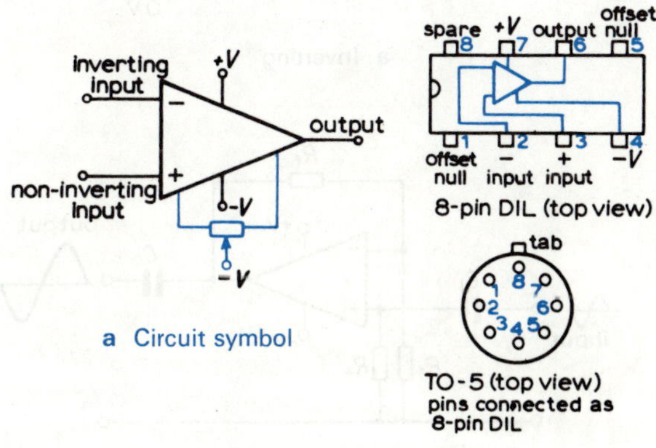

Fig. 15.3 The operational amplifier

input. Signals entering the non-inverting input are not phase shifted. Therefore the operational amplifier is able to amplify the voltage difference between the two inputs.

Two power supplies $+V$ and $-V$ are required to enable a positive and negative swing of the output voltage.

15.9 Operational amplifier as a general amplifier

Fig. 15.4a shows the additional discrete components to an operational amplifier IC required for an inverting amplifier. The input signal is presented to the inverting input with the non-inverting input clamped to the zero volt (0 V) rail.

The open-loop gain of an operational amplifier is so high that negative feedback almost always is used. Resistor R_f is the feedback resistor and the ratio of R_f to R_1 determines the closed-loop gain;

$$\text{closed-loop gain } A_{vf} = \frac{v_o}{v_{in}} = \frac{R_f}{R_1} \qquad (15.1)$$

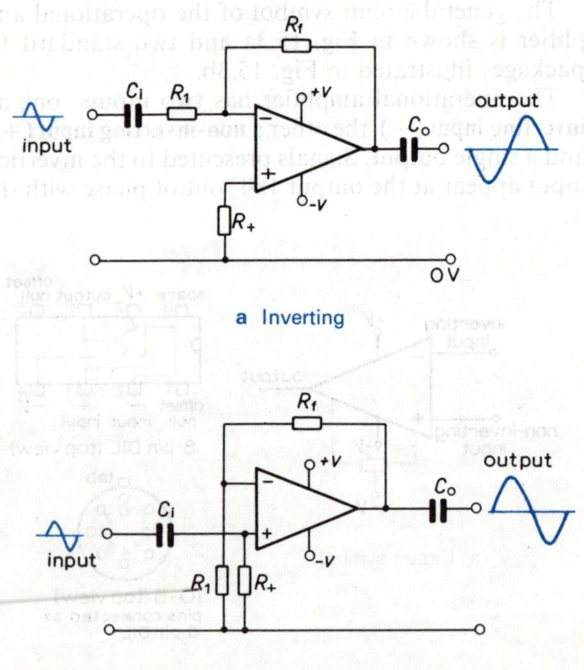

Fig. 15.4 The general operational amplifier

If a gain of 10 is required, values of $R_f = 10\ \text{M}\Omega$ and $R_1 = 1\ \text{M}\Omega$ would be typical.

The inclusion of R_+ at the non-inverting input helps prevent a drift in performance. A detailed explanation of why this is so is beyond Level III. At this stage it must suffice to define that for best performance R_+ must have the same value as R_1 in parallel with R_f;

$$R_+ = \frac{R_1 \times R_f}{R_1 + R_f} \qquad (15.2)$$

The frequency response curve of the 741 operational amplifier is shown in Fig. 15.5. (Note that the device operates from d.c. to 1 MHz because d.c., capacitors C_i and C_o are excluded.) As the frequency increases, the gain decreases at a constant rate. This is known as **roll-off**. The relationship between the frequency and maximum gain at that frequency is a constant known as the **gain-bandwidth product**. For the 741 the gain-bandwidth product is 1 MHz.

This interrelation of gain and bandwidth imposes a restriction on the use of the operational amplifier. If a gain of 100 is required, the bandwidth is limited to 10 kHz, which would not be ideal as an a.f. amplifier. If a wideband amplifier of say 100 kHz is required, the gain is limited to 10.

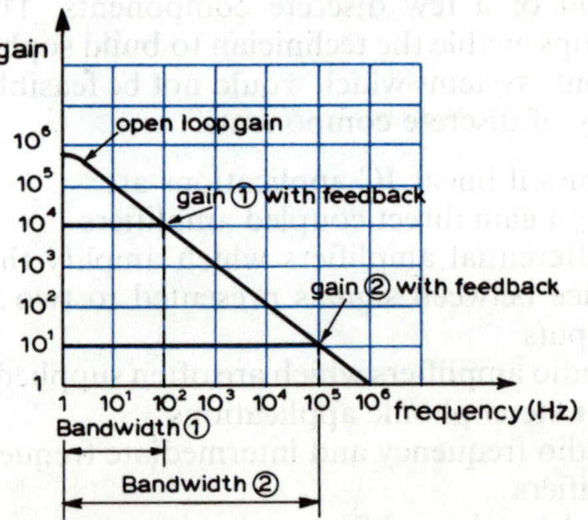

Fig. 15.5 The gain-frequency response of the 741 operational amplifier

A non-inverting amplifier is shown in Fig. 15.4b. The inverting input is clamped to the zero volt rail and the signal is presented to the non-inverting input. R_1, R_f and R_+ serve the same functions as in the inverting amplifier.

15.10 Operational amplifier as a differential amplifier

If signals are presented to both inputs of an operational amplifier through identical resistors (Fig. 15.6), the difference between the signals is amplified, the output is given by

$$v_o = A_{vf}(v_2 - v_1) \tag{15.3}$$

where v_1 is the inverting input signal and v_2 is the non-inverting input signal.

To ensure balance between the inputs, R_+ is made equivalent to R_f. The voltage gain is determined by the ratio R_f to R_1, that is

$$v_o = \frac{R_f}{R_1}(v_2 - v_1) \tag{15.4}$$

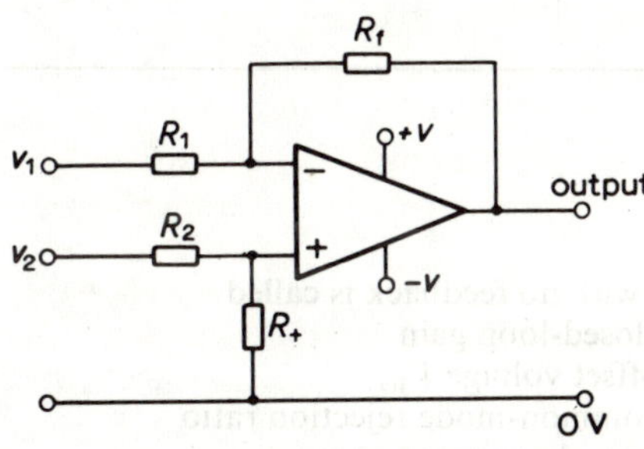

Fig. 15.6 The differential amplifier

15.11 Versatility of the operational amplifier

By connecting a variety of external components, the operational amplifier can be used as an oscillator, electronic differentiator, integrator, stabilised power supply, and so on. The examples of Fig. 15.7 illustrate the versatility of these small devices.

15.12 Selection of linear integrated circuits

The 741 operational amplifier, discussed so far, is limited to unity gain at 1 MHz, which is not ideal for many applications. Other linear operational amplifiers are specifically designed as wideband amplifiers, turned amplifiers of a.f. amplifiers. A selection of these with their main performance ratings is given in Table 15.1 on page 126.

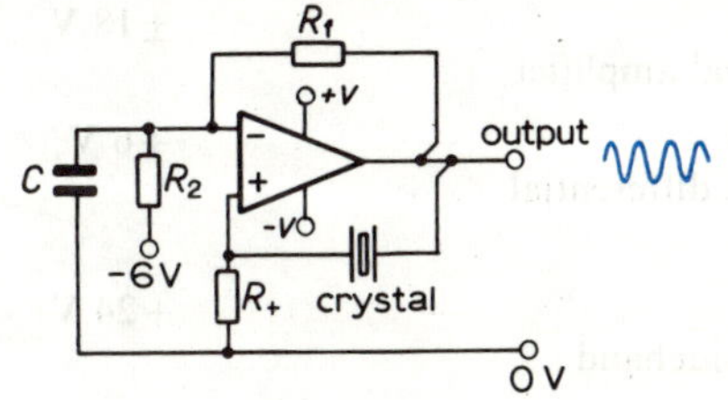

a Crystal-controlled oscillator

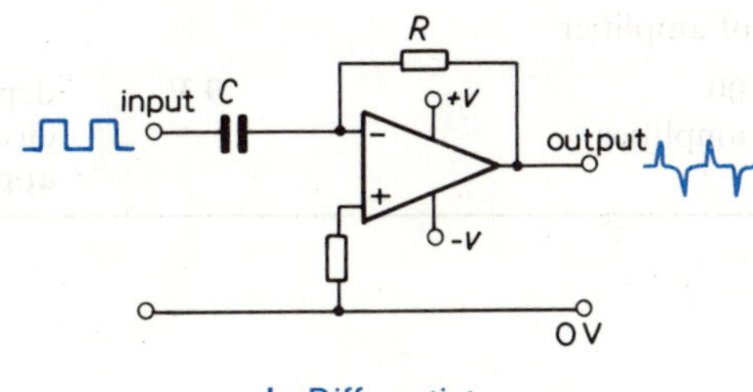

b Differentiator

c Integrator

Fig. 15.7 Uses of the operational amplifier

Review

Check your knowledge and understanding. You should be able to provide examples of the performance characteristics of currently available linear integrated circuits in relation to operational amplifiers, differential amplifiers, audio amplifiers, r.f./i.f. amplifiers, and wideband amplifiers.

Exercise

15.4 An operational amplifier is to be used as an inverting amplifier with a closed-loop gain of 30. When $R_f = 10\ \text{M}\Omega$, calculate the value of R_1, the value of R, and the bandwidth if the gain-bandwidth product is 1 MHz.

Table 15.1 Comparison of linear integrated circuits

Type	Max. Output Power	Supply Voltage	Voltage Gain	Bandwidth	Input Impedance	Output Impedance	Distortion
741 operational amplifier	—	±18 V	100 dB	Gain-bandwidth product of 1 MHz	2 MΩ	75 Ω	—
TAA 350 wideband differential amplifier	—	+6 V	80 dB	12 MHz	—	75 Ω	4%
OM 180 vhf/uhf wideband	—	+24 V	16 dB	Freq. range of 40 MHz–860 MHz	—	75 Ω	—
TAA 300 a.f. amplifier	1 W into 8 Ω	+9 V		25 kHz	15 kΩ	—	0.7%
LM 380 power of amplifier	2 W into 8 Ω	8-22 V	34 dB	100 kMz			0.2%
TAD 100 r.f./i.f. amplifier	—	9 V	depends on circuit application	—	—	—	<2%

Self assessment questions

The following are typical of questions asked in an assessment test.

Multiple choice questions

15.5 Identify which three of the following are the main advantages of circuit integration

a size
b superior performance over discrete components
c reliability
d cost
e less distortion of a signal
f greater output power possible

15.6 Identify the answer which correctly completes the following statement. The voltage gain of an operational amplifier with no feedback is called

a the closed-loop gain
b the offset voltage V_{IO}
c the common-mode rejection ratio
d the open-loop gain

Short answer questions

15.7 List typical applications of linear ICs.

15.8 Sketch and label the symbol for an operational amplifier including the inverting input, non-inverting input, positive and negative supply terminals, and output.

15.8 Describe the use of the operational amplifier as an inverting amplifier, a non-inverting amplifier, and a differential amplifier.

15.10 Explain why the 741 operational amplifier is limited as a wideband amplifier.

Appendix 1

Preferred resistance values

The following table shows the range of significant figures used for preferred resistance values. Each figure can be multiplied by any power of ten up to 10^6.

All are available with a tolerence of $\pm$ 5%. Those in bold are also available with a tolerence of $\pm$ 10%.

1.0	**1.5**	**2.2**	**2.3**	**4.7**	**6.8**
1.1	1.6	2.4	3.6	5.1	7.5
1.2	**1.8**	**2.7**	**3.9**	**5.6**	**8.2**
1.3	2.0	3.0	4.3	6.2	9.1

Appendix 2

Relationship between α and β

Current gains α and β can be related to each other for the same transistor in different configurations.

From Section 3.4:—

Eqn. 3.1 $$I_E = I_B + I_C \qquad \text{(A.2.1)}$$

therefore, $$\Delta I_E = \Delta I_B + \Delta I_C \qquad \text{(A.2.3)}$$

Now $$\alpha = \frac{\Delta I_C}{\Delta I_E} \qquad \text{(A.2.4)}$$

Subst. Eqn. A2.3 for ΔI_E into Eqn. A2.4

$$\therefore \alpha = \frac{\Delta I_C}{\Delta I_B + \Delta I_C}$$

$$\therefore \frac{1}{\alpha} = \frac{\Delta I_B + \Delta I_C}{\Delta I_C}$$

$$= \frac{\Delta I_B}{\Delta I_C} + 1$$

$$= \frac{1}{\beta} + 1 \left(\text{From } \beta = \frac{\Delta I_C}{\Delta I_B}\right)$$

$$\therefore \frac{1}{\alpha} = \frac{1 + \beta}{\beta}$$

$$\therefore \alpha = \frac{\beta}{1 + \beta} \qquad \text{(A.2.5)}$$

Rearrange for β $$\beta = \frac{\alpha}{1 - \alpha} \qquad \text{(A.2.6)}$$

Appendix 3

The gain relationship for the FET amplifier given in Chapter 9 can be derived as follows.

Consider the output of a FET amplifier (Fig. A3.1). The output can be represented by an **equivalent circuit** where the amplified signal current v_{GS} is in series with drain resistance r_D and drain load resistance R_L.

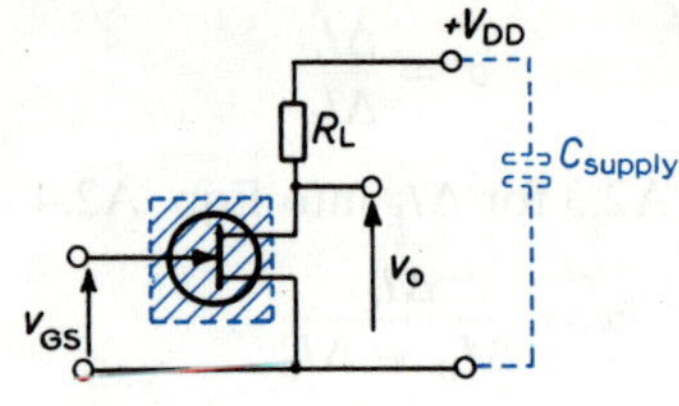

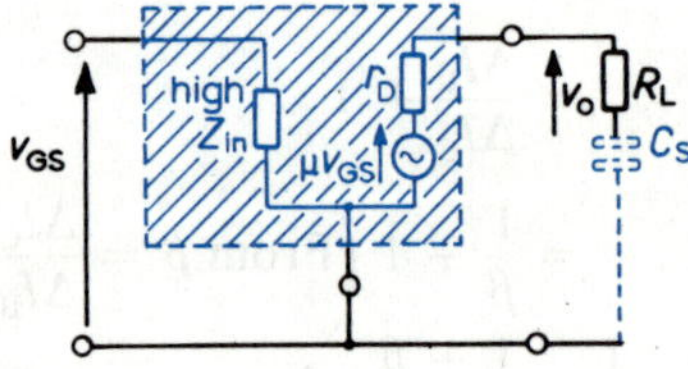

Fig. A3.1

To an a.c. signal, R_L is connected to the zero volt rail through the low capacitive reactance of the supply.

The amplified signal voltage μv_{GS} is divided between r_D and R_L

Therefore
$$v_o = \mu v_{GS} \frac{R_L}{r_D + R_L}$$

Hence the voltage gain $A_v = \dfrac{v_o}{v_{GS}} = \dfrac{\mu R_L}{r_D + R_L}$

Solutions

Chapter 1

1.1 **a** false, **b** true, **c** false
1.2 **b** and **e**
1.3 **a** 2, 8, 4 Group IV
b 2, 8, 18, 5 Group V
c 2, 8, 3 Group III
1.4 **a** true, **b** true, **c** true, **d** false, **e** false
1.5 **a**
1.6 **c**
1.7 acceptors; negative ion; *p*
1.8 **a** holes, **b** 10^{16}, **c** *p*-type
1.9 **a** *n*-type
b $2.5 \times 10^{14} + 5 \times 10^{16}$ (intrinsic electrons plus extrinsic electrons)
c 2.5×10^{14} holes
1.10 **b** and **c**
1.11 **a** true, **b** false, **c** true, **d** false
11.2 **b**
1.13 **a**

Chapter 2

2.1 **e b d a c**
2.2 **a** ≈10.6 V for Si, ≈10.2 V for Ge
b ≈0.1 V for Si, 0 V for Ge
2.3 3 A
2.4 **a** 0.44 mA, **b** 0.48 mA
2.5 forward resistance 0.833 Ω
reverse resistance 1.33 MΩ
2.6 $R_f \approx 0.26\ \Omega$; $R_r \approx 420\ \text{k}\Omega$
2.7 100 Ω (110 Ω ± 5% preferred value)
2.8 1.3 W
2.9 **a**
2.10 **b**
2.11 **a** and **d**
2.12 **b**
2.13 **b**
2.15 14.3 Ω
2.17 63.9 Ω (68 Ω ± 5% preferred value)
2.18 135.8 Ω (min) to ∞

Chapter 3

3.1 **c**
3.2 **b**
3.3 **a** 46, **b** 2.1 μA, **c** 65
3.4 **a** 49, **b** 27.6, **c** 99, **d** 39
3.5 **a**
3.6
3.7 **a** 15 kΩ, **b** 46
3.8 **a**
3.9 **b**
3.10 **a** current gain 0.98, base current 0.01 mA
b 0.77 mA
c 657 μA
d collector current 2.4 μA, base current 0.1 μA
3.11 1.8 kΩ
3.12 **b** 15.3 kΩ, **d** 99
3.13 **b** 200 kΩ, **c** 0.95

Chapter 4

4.1 **a** true, **b** false, **c** false, **d** true, **e** true
4.2 To provide correct biasing
4.3 R_L
4.4 1.08 MΩ
4.5 2.5 μA
4.6 **d**
4.8 **a** When $R_L = 1\ \text{k}\Omega$ $V_{CQ} = 4.65$ V $I_{CQ} = 3.35$ mA
b When $R_L = 1.33\ \text{k}\Omega$ $V_{CQ} = 3.7$ V $I_{CQ} = 3.25$ mA
c When $R_L = 2\ \text{k}\Omega$ $V_{CQ} = 1.9$ V $I_{CQ} = 3.1$ mA
4.9 **c**
4.10 **a** gives 150 mA peak-to-peak
4.11 **a** $A_v = 4.5$, $A_i = 4.5$
b $A_v = 4.2$, $A_i = 5.7$
4.12 quiescent points **b** to **f** inclusive
4.13 **a**
4.14 **c**
4.15 **a**
4.16 **d**
4.19 $A_i = 92$, $A_v = 23$, $A_p = 2116$

Chapter 5

5.1 **b**
5.2 **a** true, **b** false, **c** false, **d** true
5.3 **d**
5.4 **b**
5.5 **a**
5.6 **a** true, **b** false, **c** true, **d** true
5.7 **b**
5.8 **c**
5.9 **b**
5.10 **d**
5.11 **b**
5.12 **c**
5.13 **b**
5.14 **d**

Chapter 6

6.1 **a** and **e**
6.2 **b**
6.3 **a**
6.4 **a** false, **b** true, **c** true, **d** false

6.5 **d**
6.6 6
6.7 **a** Y-gain, **b** Y-shift, **c** focus, **d** time-base
6.8 amplitude 3 V, frequency 250 Hz
6.9
6.10 **c**
6.11 **d**
6.12 **b** and **d**
6.13 **d**
6.14 **a** true, **b** true, **c** false, **d** true
6.17 time-base 0.2 ms cm^{-1} (for two cycles)
Y-gain setting 20 mV cm^{-1}

Chapter 7

7.2 **a** sine/square, **b** sine/triangular
7.3 The oscillations would reduce to zero amplitude
7.4 159 kHz
7.5 **c**
7.6 **a** rectangular, **b** sawtooth, **c** rectangular, **d** sinusoidal, **e** sinusoidal
7.7 **b**
7.9 **b** and **d**
7.10 **b**

Chapter 8

8.1 **a** and **d**
8.4 **a** OR, **b** AND
8.5 AND
8.6 **a** AND **b** OR
8.7

A	B	X	Y	Z	S	C
0	0	0	1	1	0	0
0	1	1	0	0	1	0
1	0	1	0	0	1	0
1	1	1	0	1	0	1

8.8

A	B	C	X	0/P
0	0	0	0	1
0	0	1	0	1
0	1	0	0	1
0	1	1	0	1
1	0	0	0	1
1	0	1	0	1
1	1	0	0	1
1	1	1	1	0

This is the NOT AND or NAND function.
8.9 **a, d, f**
8.11 **b**
8.12 **b**
8.13 **b**

Chapter 9

9.1 **c**
9.2 JUGFET (*n*-channel)
9.3 **b**
9.4 **a** false, **b** true, **c** false, **d** true
9.5 **c**
9.6 enhancement
9.7 **a, c** and **e**
9.8 **a** 180, **b** 0.595 mS
9.9 9.2
9.10 **bi** 10 kΩ, **ii** 2.2 mS, **iii** 220 **c** 4.3
9.11 **a** true, **b** false, **c** false, **d** true, **e** false
9.12 IGFET
9.13 **b**
9.14 depletion
9.15 **a**
9.16 **a**
9.17 **a-i, b-v, c-ii**
9.18 **c**
9.22 8.5

Chapter 10

10.1 **d**
10.2 **b, c, d, g, h**
10.3 **d**
10.4 **a** true, **b** false, **c** false, **d** true, **e** true, **f** true
10.5 **b**
10.6 **b**
10.7 **d**
10.8 **a** true, **b** true, **c** true, **d** false, **e** true, **f** false
10.9 **d**
10.10 **b**
10.11 **c**
10.12 Class A
10.13 **b**
10.14 **a iii** and **iv**, **b ii**
10.15 **a** false, **b** false, **c** false, **d** true, **e** true
10.16 **d**
10.17 **c**
10.18 **a**
10.19 **b**

Chapter 11

11.1 **a** true, **b** false, **c** true, **d** false, **e** false
11.2 **a** 18.2 μV, **b** 0, **c** 9.2 μV, **d** 20.3 μV
11.3 32.2 dB
11.4 58 dB
11.5 **a** 10^{-4} W, **b** 25.3 mV
11.6 **a** internal, **b** external, **c** internal, **d** internal, **e** external, **f** external, **g** internal, **h** internal
11.7 **a** flicker, **b** Johnson's or white, **c** microphonic, **d** flicker, **e** shot
11.8 **b**
11.11 31.25 dB

Chapter 12

12.1 **a** false, **b** false, **c** false, **d** true, **e** true
12.2 larger
12.3 **b**
12.5 **d**
12.6 A_{vf} approaches infinity, limited only by supply voltage
12.7 **a i** 99, **ii** 0.049%, **iii** 2020 kHz **iv** 50 dB
b i 33.2, **ii** 0.017%, **iii** 6020 kHz, **iv** 54.8 dB
12.8 **d**
12.9 **a** the voltage, **b** in series, **c iii**
12.10 **a-vi, b-iv** or **ii, c-ii, d-iii, e-iii, f-i**
12.11 **b** and **c**
12.12 **a**

Chapter 13
13.1 a
13.2 a false, **b** true, **c** false, **d** true
13.3 b
13.4 a true, **b** false, **c** false, **d** true
13.7

Chapter 14
14.1 a bistable, **b** monostable, **c** astable
14.2 a false, **b** true, **c** true, **d** false, **e** false
14.3 b
14.4 4:1
14.5 c
14.6 b
14.7 b
14.8 d
14.9 a bistable, **b** astable, **c** monostable
14.10 b
14.11 d
14.12 a
14.13 a
14.14 a
14.15 differentiating circuit

Chapter 15
15.1 c
15.2 c
15.3 a DIL, **b** TO-5
15.4 a 333 kΩ, **b** 248 kΩ, **c** 33 kHz
15.5 a, c and d
15.6 d

Index